New simple ways to solve equations

NEW SIMPLE WAYS TO SOLVE

EQUATIONS

How to solve equations by mental arithmetic, which strengthens
the ability to think and improves the memory

EINAR ÖSTMYREN

ISBN: 978-91-7699-768-0

Copyright © 2018 Einar Östmyren

First edition: March 2018

Second edition: March 2022

Design: Stefan "Lillis" Åkesson

Publishing firm: BoD – Books on Demand, Stockholm, Sweden

Printing: BoD – Books on Demand, Norderstedt, Germany

CONTENTS

INTRODUCTION

In this book I present a unique quadratic formula, which turned out to be a rewriting of the p-q-formula. This rewriting resulted in the equations being solved, almost twice as fast with the new formula when it was compared in a test with the p-q-formula. Another test also showed that the new formula was much faster than the Vedic formula, see p.11. The new formula is unique because the equations in the test were solved by mere mental calculation, which improves the memory and increases mental agility and intelligence.

When I discovered that the middle coefficient in a quadratic equation contains all information about its origin, it led to a rule that simplified the solving of all equations. In a quadratic equation the origin could be located, and it became then possible to create a rule for how the coefficients were to be split up into factors. By means of this rule and some exercises the answer to an equation can both be calculated and checked regardless how large the coefficients are. This universal method should be used before the equation is solved by a formula, see pages 20-23.

Since the origin of a quadratic equation could be located, it was also simple to find the origin to other types of equations, and therefore new methods could be created. This led to the fact that a cubic equation could be solved without taking detours, like polynomial division, a guess or a test of a root. When the origin of an equation can be located, it is as easy to solve a fifth degree equation as a quadratic equation, in the same simple way as unlocking a safe with a key. The purpose of the book is mainly to make it as easy as possible for the students to solve equations, but also to give them a good insight into the origin of an equation.

During a visit to USA I met Dr. Anne Dow, Professor of Mathematics at MUM in Iowa, and the new quadratic formula was then verified and approved.

I was born in 1932 in the Norwegian city of Risör, and when I was 21years old I studied in Sweden to become a Chemical Engineer.

Chapter 1

THE NEW QUADRATIC FORMULA

Using the new formula we have seen that the equations can be solved mentally, and also almost twice as fast as by the p-q-formula. The result of a comparative test of the two formulas and a Vedic formula is accounted on page 18.

I had long tried to find a simple formula, but when I could solve several equations with irrational numbers, I realized that I had succeeded. Later it turned out that the formula is a rewrite of the p-q-formula.

The formulas are equal but have different properties. They can be compared to two equal cars, one of them with a stronger motor and almost twice as fast as the other one.

Derivation of the new quadratic formula

The new formula: $Ax = -0.5B \pm \sqrt{(0.5B)^2 + A(-C)}$

We want to show that $Ax^2 + Bx + C = 0$ is equivalent to the new quadratic formula.

$Ax^2 + Bx + C = 0$ multiply both sides by A

$A^2x^2 + ABx + AC = 0$ complement both sides with $\left(\frac{B}{2}\right)^2$

$A^2x^2 + ABx + \left(\frac{B}{2}\right)^2 = \left(\frac{B}{2}\right)^2 - AC$ the squaring rule gives:

$\left(Ax + \frac{B}{2}\right)^2 = \left(\frac{B}{2}\right)^2 - AC$

$Ax + \frac{B}{2} = \pm\sqrt{\left(\frac{B}{2}\right)^2 - AC}$

$Ax = -\frac{B}{2} \pm \sqrt{\left(\frac{B}{2}\right)^2 - AC}$

$Ax = -0.5B \pm \sqrt{(0.5B)^2 + A(-C)}$ *

$Ax^2 + Bx + C = 0$ is equivalent to $Ax = -0.5B \pm \sqrt{(0.5B)^2 + A(-C)}$.

* Squaring of numbers ending in 5, see page 9.

The conjugate rule

The new formula for quadratic equations is based on the conjugate rule, which is often used to rewrite a difference into a product. If a and b are two numbers we have:

$$(a + b)(a - b) = a^2 - b^2$$

This identity is valid for any numbers a and b. The conjugate rule can be rewritten into a product, where one factor has plus and the other one has minus between its parentheses.

$$a^2 - b^2 = (a + b)(a - b)$$

When multiplying for example $27 \cdot 33$, the average is 30 and the difference is 3. When a $=30$ and b $= 3$ we get $(30 + 3)(30 - 3) = 30^2 - 3^2 = 900 - 9 = 891$, i.e. $27 \cdot 33 = 891$.

The conjugate rule can often be used for swift and elegant solutions, which is illustrated in the following examples.

Example 1 $71^2 - 69^2$ Difference between two squares

$$71^2 - 69^2 = (71 + 69)(71 - 69) = 140 \text{ x } 2 = 280$$

Example 2 $x^2 = 29^2 - 21^2$ Pythagoras´s theorem

$$x^2 = (29 + 21)(29 - 21) = 50 \text{ x } 8 = 400$$
$$x = 20$$

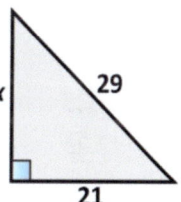

Squaring of numbers ending in 5

As the middle coefficient is halved in the new formula, odd numbers will end in 5. An easy way to square such numbers by mental calculation, is to use a word formula : " By one more than the one before". By squaring of 5.5, 5 is multiplied by 6, i.e. 5 is multiplied by the number following 5. We multiply 5 by 6 and get 30, to which we add 25 in appropriate decimal position.

Example: Multiply $5 + 0.5$ by $6 - 0.5 = 30.25$ and $5.5^2 = 30.25$.

Solve the equation: $4x^2 + 15x + 9 = 0$

The middle coefficient 15 is halved to 7.5. To square 7.5 we multiply 7 by 8, which is 56 and then we add 25 in the decimal position where it belongs, so $7.5^2 = 56.25$.

$$Ax = -0.5B \pm \sqrt{(0.5B)^2 + A(-C)}$$

$$4x = -7.5 \pm \sqrt{(56.25 + 4(-9))} = -7.5 \pm \sqrt{20.25} = -7.5 \pm 4.5$$

$$x_1 = -3, \; x_2 = -3/4$$

Table for squaring and square roots

$0.5^2 = 0.25$	$\sqrt{0{,}25} = 0.5$	$1^2 = 1$	$\sqrt{1} = 1$
$1.5^2 = 2.25$	$\sqrt{2{,}25} = 1.5$	$2^2 = 4$	$\sqrt{4} = 2$
$2.5^2 = 6.25$	$\sqrt{6{,}25} = 2.5$	$3^2 = 9$	$\sqrt{9} = 3$
$3.5^2 = 12.25$	$\sqrt{12{,}25} = 3.5$	$4^2 = 16$	$\sqrt{16} = 4$
$4.5^2 = 20.25$	$\sqrt{20{,}25} = 4.5$	$5^2 = 25$	$\sqrt{25} = 5$
$5.5^2 = 30.25$	$\sqrt{30{,}25} = 5.5$	$6^2 = 36$	$\sqrt{36} = 6$
$6.5^2 = 42.25$	$\sqrt{42{,}25} = 6.5$	$7^2 = 49$	$\sqrt{49} = 7$
$7.5^2 = 56.25$	$\sqrt{56{,}25} = 7.5$	$8^2 = 64$	$\sqrt{64} = 8$
$8.5^2 = 72.25$	$\sqrt{72{,}25} = 8.5$	$9^2 = 81$	$\sqrt{81} = 9$
$9.5^2 = 90.25$	$\sqrt{90{,}25} = 9.5$	$10^2 = 100$	$\sqrt{100} = 10$
$10.5^2 = 110.25$	$\sqrt{110{,}25} = 10.5$	$11^2 = 121$	$\sqrt{121} = 11$
$11.5^2 = 132.25$	$\sqrt{132{,}25} = 11.5$	$12^2 = 144$	$\sqrt{144} = 12$

This table contains numbers which are often needed to solve quadratic equations by the new formula. If you are familiar with the small multiplication table, it should not be any problem to square and calculate the square root by mental calculation. Let us determine the square root of 12.25. We realize that the number is more than 3 and less than 4. It is between 3 and 4, i.e. 3.5. The square root of 72.25 should be between 8 and 9 and can be verified to be 8.5. With some exercise you will soon solve equations faster by mental calculation than by means of a technical device.

The new formula is tested by students

A test was performed by the students at Blackeberg Gymnasium, a senior high school, which is ranked as one of the best schools in Stockholm, Sweden. Before the students received the exercises a comparison was made, which turned out that the equations were solved faster by the Vedic formula than by the p-q-formula. Therefore the new formula was only compared to the Vedic formula in the test performed by the students. The test was performed by 21 students and consisted of 20 equations.

To show how cumbersome and time-consuming the p-q-formula is compared with the new formula, each equation was solved by the p-q-formula.

Formula A: $Ax = -0.5B \pm \sqrt{(0.5B)^2 + A(-C)}$ the new formula

Formula B: $2Ax = -B \pm \sqrt{(B)^2 - 4AC}$ the Vedic formula

Formula C: $x = -\frac{p}{2} \pm \sqrt{\left(\frac{p}{2}\right)^2 - q}$ the p-q-formula

Exercises during the test

Example 1 $2x^2 - 16x + 14 = 0$ $\qquad\qquad$ $x_1 = 1,\;\; x_2 = 7$

$\quad$ A: $2x = 8 \pm \sqrt{64 + 2(-14)} = 8 \pm\sqrt{36} = 8 \pm 6$

$\quad$ B: $4x = 16 \pm \sqrt{256 - 4 \cdot 2 \cdot 14} = 16 \pm\sqrt{144} = 16 \pm 12$

$\quad$ C: $x^2 + \frac{16x}{2} + \frac{14}{2} = 0,\; x = \frac{16}{4} \pm \sqrt{\frac{256-112}{16}} = \frac{16}{4} \pm \sqrt{\frac{144}{16}} = \frac{16}{4} \pm \frac{12}{4}$

Example of the graph to the quadratic function: $y = 2x^2 - 16x + 14$

a)

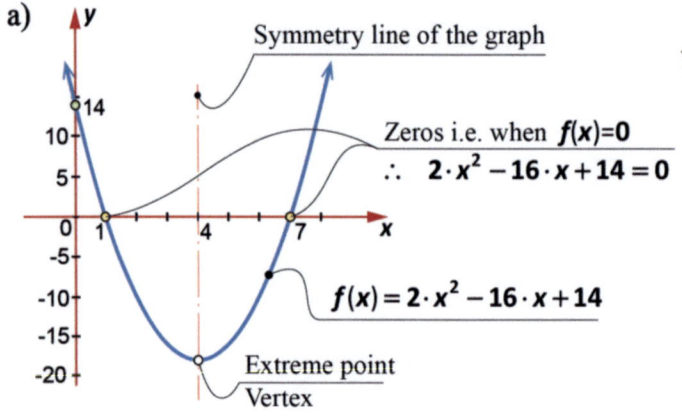

Symmetry line of the graph

Zeros i.e. when $f(x)=0$

$\therefore$ $\;2\cdot x^2 - 16\cdot x + 14 = 0$

$f(x) = 2\cdot x^2 - 16 \cdot x + 14$

Extreme point
Vertex

b)

x	$f(x)$	(x, y)
0	14	0, 14
1	0	1, 0
2	-10	2, -10
3	-16	3, -16
4	-18	4, -18
5	-16	5, -16
6	-10	6, -10
7	0	7, 0
8	14	8, 14

Example 2 $2x^2 + 17x + 8 = 0$ $\qquad\qquad$ $x_1 = -8,\;\; x_2 = -1/2$

$\quad$ A: $2x = -8.5 \pm\sqrt{72.25 + 2(-8)} = -8.5 \pm\sqrt{56.25} = -8.5 \pm 7.5$

$\quad$ B: $4x = -17\pm\sqrt{289 - 4 \cdot 2 \cdot 8} = -17 \pm\sqrt{225} = -17 \pm 15$

$\quad$ C: $x^2 + \frac{17x}{2} + \frac{8}{2} = 0\; x = -\frac{17}{4} \pm \sqrt{\frac{289-64}{16}} = -\frac{17}{4} \pm \sqrt{\frac{225}{16}} = -\frac{17}{4} \pm \frac{15}{4}$

Example 3 $2x^2 + 7x + 5 = 0$ $\qquad\qquad x_1 = -1, \quad x_2 = -5/2$

A: $2x = -3.5 \pm \sqrt{12.25 + 2(-5)} = -3.5 \pm \sqrt{2.25} = -3.5 \pm 1.5$

B: $4x = -7 \pm \sqrt{49 - 4 \cdot 2 \cdot 5} = -7 \pm \sqrt{9} = -7 \pm 3$

C: $x^2 + \frac{7x}{2} + \frac{5}{2} = 0$ $x = -\frac{7}{4} \pm \sqrt{\frac{49-40}{16}} = -\frac{7}{4} \pm \sqrt{\frac{9}{16}} = -\frac{7}{4} \pm \frac{3}{4}$

Example 4 $3x^2 + 14x + 8 = 0$ $\qquad\qquad x_1 = -4, \quad x_2 = -2/3$

A: $3x = -7 \pm \sqrt{49 + 3(-8)} = -7 \pm \sqrt{25} = -7 \pm 5$

B: $6x = -14 \pm \sqrt{196 - 4 \cdot 3 \cdot 8} = -14 \pm \sqrt{100} = -14 \pm 10$

C: $x^2 + \frac{14x}{3} + \frac{8}{3} = 0$ $x = -\frac{14}{6} \pm \sqrt{\frac{196-96}{36}} = -\frac{14}{6} \pm \sqrt{\frac{100}{36}} = -\frac{14}{6} \pm \frac{10}{6}$

Example 5 $4x^2 + 18x + 8 = 0$ $\qquad\qquad x_1 = -4, \quad x_2 = -1/2$

A: $4x = -9 \pm \sqrt{81 + 4(-8)} = -9 \pm \sqrt{49} = -9 \pm 7$

B: $8x = -18 \pm \sqrt{324 - 4 \cdot 4 \cdot 8} = -18 \pm \sqrt{196} = -18 \pm 14$

C: $x^2 + \frac{18x}{4} + \frac{8}{4} = 0$ $x = -\frac{18}{8} \pm \sqrt{\frac{324-128}{64}} = -\frac{18}{8} \pm \sqrt{\frac{196}{64}} = -\frac{18}{8} \pm \frac{14}{8}$

Example 6 $x^2 + 11x + 18 = 0$ $\qquad\qquad x_1 = -2, \quad x_2 = -9$

A: $x = -5.5 \pm \sqrt{30.25 + 1(-18)} = -5.5 \pm \sqrt{12.25} = -5.5 \pm 3.5$

B: $2x = -11 \pm \sqrt{121 - 4 \cdot 1 \cdot 18} = -11 \pm \sqrt{49} = -11 \pm 7$

C: $x = -\frac{11}{2} \pm \sqrt{\frac{121-72}{4}} = -\frac{11}{2} \pm \sqrt{\frac{49}{4}} = -\frac{11}{2} \pm \frac{7}{2}$

Example 7 $8x^2 + 2x - 15 = 0$ $x_1 = 5/4,\quad x_2 = -3/2$

A: $8x = -1 \pm \sqrt{1 + 8 \cdot 15} = -1 \pm \sqrt{121} = -1 \pm 11$

B: $16x = -2 \pm \sqrt{4 - 4 \cdot 8(-15)} = -2 \pm \sqrt{484} = -2 \pm 22$

C: $x^2 + \dfrac{2x}{8} - \dfrac{15}{8} = 0,\ x = -\dfrac{2}{16} \pm \sqrt{\dfrac{4 + 480}{256}} = -\dfrac{2}{16} \pm \sqrt{\dfrac{484}{256}} = -\dfrac{2}{16} \pm \dfrac{22}{16}$

Example 8 $4x^2 + 15x + 9 = 0$ $x_1 = -3,\ x_2 = -3/4$

A: $4x = -7.5 \pm \sqrt{56.25 + 4(-9)} = -7.5 \pm \sqrt{20.25} = -7.5 \pm 4.5$

B: $8x = -15 \pm \sqrt{225 - 4 \cdot 4 \cdot 9} = -15 \pm \sqrt{81} = -15 \pm 9$

C: $x^2 + \dfrac{15x}{4} + \dfrac{9}{4} = 0,\ x = -\dfrac{15}{8} \pm \sqrt{\dfrac{225 - 144}{64}} = -\dfrac{15}{8} \pm \sqrt{\dfrac{81}{64}} = -\dfrac{15}{8} \pm \dfrac{9}{8}$

Example 9 $5x^2 - 9x - 18 = 0$ $x_1 = 3,\quad x_2 = -6/5$

A: $5x = 4.5 \pm \sqrt{20.25 + 5 \cdot 18} = 4.5 \pm \sqrt{110.25} = 4.5 \pm 10.5$

B: $10x = 9 \pm \sqrt{81 - 4 \cdot 5(-18)} = 9 \pm \sqrt{441} = 9 \pm 21$

C: $x^2 - \dfrac{9x}{5} - \dfrac{18}{5} = 0,\ x = \dfrac{9}{10} \pm \sqrt{\dfrac{81 + 360}{100}} = \dfrac{9}{10} \pm \sqrt{\dfrac{441}{100}} = \dfrac{9}{10} \pm \dfrac{21}{10}$

Example 10 $x^2 - 6x + 2 = 0$ $x_1 = 3 + \sqrt{7},\ x_2 = 3 - \sqrt{7}$

A: $x = 3 \pm \sqrt{9 + 1(-2)} = 3 \pm \sqrt{7}$

B: $2x = 6 \pm \sqrt{36 - 4 \cdot 1 \cdot 2} = \dfrac{6}{2} \pm \dfrac{\sqrt{28}}{2} = 3 \pm \sqrt{7}$

C: $x = \dfrac{6}{2} \pm \sqrt{\dfrac{36 - 8}{4}} = \dfrac{6}{2} \pm \sqrt{\dfrac{28}{4}} = \dfrac{6}{2} \pm \dfrac{\sqrt{28}}{2} = 3 \pm \sqrt{7}$

Example 11 $2x^2 - 4x + 5 = 0$ $\qquad\qquad$ $x_1 = 1 + \dfrac{i\sqrt6}{2}, \quad x_2 = 1 - \dfrac{i\sqrt6}{2}$

A: $2x = 2 \pm \sqrt{4 + 2(-5)} = 1 \pm \dfrac{i\sqrt6}{2}$

B: $4x = 4 \pm \sqrt{16 - 4 \cdot 2 \cdot 5} = 1 \pm \dfrac{i\sqrt{24}}{4} = 1 \pm \dfrac{i\sqrt6}{2}$

C: $x^2 = -\dfrac{4}{2} + \dfrac{5}{2} = 0,\ x = \dfrac{4}{4} \pm \sqrt{\dfrac{16-40}{16}} = 1 \pm \dfrac{i\sqrt{24}}{4} = 1 \pm \dfrac{i\sqrt6}{2}$

Example 12 $3x^2 + 13x + 12 = 0$ $\qquad\qquad$ $x_1 = -3, \quad x_2 = -4/3$

A: $3x = -6.5 \pm \sqrt{42.25 + 3(-12)} = -6.5 \pm \sqrt{6.25} = -6.5 \pm 2.5$

B: $6x = -13 \pm \sqrt{169 - 4 \cdot 3 \cdot 12} = -13 \pm \sqrt{25} = -13 \pm 5$

C: $x^2 + \dfrac{13x}{3} + \dfrac{12}{3} = 0,\ x = -\dfrac{13}{6} \pm \sqrt{\dfrac{169-144}{36}} = -\dfrac{13}{6} \pm \sqrt{\dfrac{25}{36}} = -\dfrac{13}{6} \pm \dfrac{5}{6}$

Example 13 $x^2 - 18x + 81 = 0$ $\qquad\qquad$ $x_1 = x_2 = 9$

A: $x = 9 \pm \sqrt{81 + 1(-81)} = 9$

B: $2x = 18 \pm \sqrt{324 - 4 \cdot 1 \cdot 81} = 18$

C: $x = \dfrac{18}{2} \pm \sqrt{\dfrac{324-324}{4}} = 9$

Example 14 $x^2 - 10x + 21 = 0$ $\qquad\qquad$ $x_1 = 7, \quad x_2 = 3$

A: $x = 5 \pm \sqrt{25 + 1(-21)} = 5 \pm \sqrt4 = 5 \pm 2$

B: $2x = 10 \pm \sqrt{100 - 4 \cdot 1 \cdot 21} = 10 \pm \sqrt{16} = 10 \pm 4$

C: $x = \dfrac{10}{2} \pm \sqrt{\dfrac{100-84}{4}} = \dfrac{10}{2} \pm \sqrt{\dfrac{16}{4}} = \dfrac{10}{2} \pm \dfrac{4}{2}$

Example 15 $3x^2 + 5x - 8 = 0$ $x_1 = -8/3, \quad x_2 = 1$

A: $3x = -2.5 \pm \sqrt{6.25 + 3 \cdot 8} = -2.5 \pm \sqrt{30.25} = -2.5 \pm 5.5$

B: $6x = -5 \pm \sqrt{25 - 4 \cdot 3(-8)} = -5 \pm \sqrt{121} = -5 \pm 11$

C: $x^2 + \frac{5x}{3} - \frac{8}{3} = 0, \; x = -\frac{5}{6} \pm \sqrt{\frac{25+96}{36}} = -\frac{5}{6} \pm \sqrt{\frac{121}{36}} = -\frac{5}{6} \pm \frac{11}{6}$

Example 16 $3x^2 - 10x - 8 = 0$ $x_1 = 4, \quad x_2 = -2/3$

A: $3x = 5 \pm \sqrt{25 + 3 \cdot 8} = 5 \pm \sqrt{49} = 5 \pm 7$

B: $6x = 10 \pm \sqrt{100 - 4 \cdot 3(-8)} = 10 \pm \sqrt{196} = 10 \pm 14$

C: $x^2 + \frac{10x}{3} + \frac{8}{3} = 0 \; x = -\frac{10}{6} \pm \sqrt{\frac{100+96}{36}} = \frac{10}{6} \pm \sqrt{\frac{196}{36}} = \frac{10}{6} \pm \frac{14}{6}$

Example 17 $3x^2 + 2x - 4 = 0$ $x_1 = -\frac{1}{3} + \frac{\sqrt{13}}{3}, \quad x_2 = -\frac{1}{3} - \frac{\sqrt{13}}{3}$

A: $3x = -1 \pm \sqrt{1 + 3 \cdot 4} = -\frac{1}{3} \pm \frac{\sqrt{13}}{3}$

B: $6x = -2 \pm \sqrt{4 - 4 \cdot 3(-4)} = -\frac{2}{6} \pm \frac{\sqrt{52}}{6} = -\frac{1}{3} \pm \frac{\sqrt{13}}{3}$

C: $x^2 + \frac{2x}{3} - \frac{4}{3} = 0, \; x = -\frac{2}{6} \pm \sqrt{\frac{4+48}{36}} = -\frac{2}{6} \pm \frac{\sqrt{52}}{6} = -\frac{1}{3} \pm \frac{\sqrt{13}}{3}$

Example 18 $x^2 + 5x + 6 = 0$ $x_2 = -3, \quad x_1 = -2$

A: $x = -2.5 \pm \sqrt{6.25 + 1(-6)} = -2.5 \pm \sqrt{0.25} = -2.5 \pm 0.5$

B: $2x = -5 \pm \sqrt{25 - 4 \cdot 1 \cdot 6} = -5 \pm \sqrt{1} = -5 \pm 1$

C: $x = -\frac{5}{2} \pm \sqrt{\frac{25-24}{4}} = -\frac{5}{2} \pm \sqrt{\frac{1}{4}} = -\frac{5}{2} \pm \frac{1}{2}$

Example 19 $x^2 + 4x - 3 = 0$ $\qquad\qquad$ $x_1 = -2+\sqrt{7}, \quad x_2 = -2-\sqrt{7}$

A: $x = -2 \pm \sqrt{4 + 1 \cdot 3} = -2\pm\sqrt{7}$

B: $2x = -4 \pm \sqrt{16 - 4 \cdot 1(-3)} = -\dfrac{4}{2} \pm \dfrac{\sqrt{28}}{2} = -2\pm\sqrt{7}$

C: $x = -\dfrac{4}{2} \pm \sqrt{\dfrac{16+12}{4}} = -\dfrac{4}{2} \pm \sqrt{\dfrac{28}{4}} = -\dfrac{4}{2} \pm \dfrac{\sqrt{28}}{2} = -2\pm\sqrt{7}$

Example 20 $6x^2 + 17x + 12 = 0$ $\qquad\qquad$ $x_1 = -3/2, \quad x_2 = -4/3$

A: $6x = -8.5 \pm \sqrt{72.25 + 6(-12)} = -8.5 \pm\sqrt{0.25} = -8.5 \pm 0.5$

B: $12x = -17 \pm \sqrt{289 - 4 \cdot 6 \cdot 12} = -17 \pm\sqrt{1} = -17 \pm 1$

C: $x^2 + \dfrac{17x}{6} + \dfrac{12}{6} = 0, \quad x = -\dfrac{17}{12} \pm \sqrt{\dfrac{289-288}{144}} = -\dfrac{17}{12} \pm \sqrt{\dfrac{1}{144}} = -\dfrac{17}{12} \pm \dfrac{1}{12}$

The result of the test

After the test all the students liked the new formula, because it was so simple to square the small numbers by mental calculation. As the students were not used to the formulas, the time for solving the equations varied. Therefore it was not easy to evaluate the result of the test but, on average the students solved the equations faster with the new formula than with the Vedic formula. A few students solved the equations on average 16% faster with the new formula than with the Vedic formula.

When the same test was performed later by two experienced persons, all equations were solved by mental calculation when they used the new formula. The result of the test was that they solved the equations on average 30 % faster with the new formula compared with the Vedic formula and 75 % faster compared with the p-q-formula.

The new formula has shown to be the best, and is also able to solve almost all equations by mental calculation.

In favour of mental mathematics

The following points outline the benefits of a mental approach to mathematics.

1. Mental calculation sharpens the mind and increases intelligence and mental agility. This will be evident to anyone who has practiced mental calculation or who has seen its effect.

2. Mental calculation enhances the precision of thought. Numbers and other mathematical objects are neutral, giving only one correct answer to which everyone will agree and there is never a contradiction. This absolute precision is unique to mathematics so dealing intimately with numbers as we do by mental calculation, we cultivate fine and careful thinking.

3. Mental calculation leads naturally to the search for, and discernment of, constancy and law, which are very necessary attributes in a swiftly changing world.

4. Our mind has the ability to retain several ideas at once so that they can be compared and combined. This facility is enhanced by mental calculation as we practice holding the sum in the mind whilst operating with some of the figures.

5. Mental calculation improves the memory and depreciates if it is not exercised. Short term, medium term and long term memory are all stimulated by mental calculation.

6. Since numbers are absolutely reliable, mental calculation promotes confidence. In particular mental calculation creates confidence in oneself and in one's capabilities. To solve problems, a difficult one, by mental arithmetic without having to rely on artificial aid is a source of great satisfaction and encouragement.

7. Mental calculation is a delight to the mind: the intrinsic qualities, relationships and beauty of numbers, the way they create new numbers out of themselves is a source of great enjoyment.

8. Through mental calculation one becomes familiar with numbers and appreciates their various properties. This leads to real understanding of numbers.

9. Calculating mentally reveals subtle properties of numbers and their relationships more readily than if the calculation was written down and thereby fixed. Thus mental calculation leads naturally to innovation and to the invention of new methods, thereby developing the student's natural creativity.

10. Practical uses of mental calculation are many, as we all need to make quick on the spot, calculations from time to time.

To sum up we can see that mental calculation has many advantages and really brings mathematics to life as well as it provides motivation, strengthening and enlivening the mind. Our mind operates fast and has a variety of operational properties. By proper training we can use these properties of the mind to our advantage.

This is not say that pencil and paper or calculator are to be avoided in mathematics, they have of course their place but, mental calculation should be the primary method of calculation.

Chapter 2

QUADRATIC EQUATIONS

Factorising quadratic equations is based on vertically and crosswise multiplication, and a method how to split up the coefficients into factors. As the origin of the equation can be located, the answer to an equation can be calculated and checked before the equation is solved by a formula. In addition to the exercises on pages 22-23 you can also practice the examples on page 41. With some practice many equations can be solved by mental calculation and it will give you increased self-confidence, satisfaction and joy.

A polynomial is an algebraic expression for a sum of terms. A term consists of a coefficient and an integer power of a variable. In the polynomial $2x^2 + 16x + 14$, x is a variable, 2 and 16 coefficients and 14 is a constant term. When multiplying two numbers you get a product, but factorising is a reverse operation, where you start from a product and split it into factors.

Let us take a simple example and factorise $12 = 3 \cdot 4$, where 3 and 4 are factors. We have split 12 into 3 and 4, and in the same way we can split a polynomial, e.g. $x^2 + 4x$. We have two terms both containing x. When $x^2 + 4x$ is factorised we get two factors, x and $(x + 4)$.

Crosswise multiplication of algebraic expressions

Here is a general method of multiplication, by which algebraic expressions can be multiplied in one line and from left to right or from right to left.

1st step   2nd step   3rd step

A quadratic polynomial is constructed with the binomials $5x + 2$, $x + 4$. Put the two binomials under each other and perform vertically and crosswise multiplication as shown below.

$5x + 2$
$x + 4 = 5x^2 + 22x + 8$
1st step: $5x \cdot x = 5x^2$
2nd step: $5x \cdot 4 + x \cdot 2 = 22x$
3rd step : $2 \cdot 4 = 8$

The middle coefficient 22 is the sum of two products and it is important when you split up the coefficient into factors.

Table how to split up the coefficients into factors

The table shows how the coefficients are split up into factors. The prime numbers 2, 3, 5, 7, 11, 13, etc. can only be split up into themselves and 1 while others can be split into a different way as shown below.

1. 1/1	2. 2/1	3. 3/1	4. 4/1, 2/2
5. 5/1	6. 2/3, 1/6	7. 7/1	8. 2/4, 1/8
9. 3/3, 1/9	10. 2/5, 1/10	11. 11/1	12. 3/4, 2/6, 1/12
13. 13/1	14. 2/7, 1/14	15. 3/5, 1/15	16. 2/8, 4/4, 1/16
17. 17/1	18. 3/6, 2/9, 1/18	19. 19/1	20. 4/5, 2/10, 1/20

The split up rule (It means how the coefficients are to be split up into factors).

If the middle coefficient is an odd number, one odd factor in the first term is multiplied crosswise by an odd factor in the constant term. The other two factors can either be even factors or an odd and an even factor.

If the middle coefficient is an even number 4 odd factors, 4 even factors or 2 even and 2 odd factors have been mixed in pairs and multiplied crosswise e.g. $3 \cdot 4 + 1 \cdot 2 = 14$.

Let us illustrate this with an example, $6x^2 + 17x + 12 = 0$

The left side of the equation is factorised and since the middle coefficient 17 is an odd number, an odd factor in the first term has to be cross-multiplied by an odd factor in the last term. It means that $6x^2$ is split into $3x/2x$ and 12 is split into 4/3. Note that two odd factors 3x and 3 have been cross-multiplied as follows:

3x + 4 $\qquad\qquad\qquad\qquad x_1 = -4/3, \; x_2 = -3/2$

$2x + \mathbf{3} = 6x^2 + 17x + 12$

If you know how to split up the coefficients into factors and have some experience, the answer to an equation can be calculated within a few seconds. After cross-multiplication of the binomials the result has to be identical with the polynomial.

Good to remember

1. Minus x minus = plus, minus x plus = minus, plus x plus = plus.
2. To simplify we often write the first and the last coefficient when we really mean the first coefficient and the constant term.
3. Please check first if the middle coefficient is an odd or an even number.
4. A binomial consists of two terms, e.g. $3x + 1$.
5. When we write e.g. $3x \cdot 3$ it means crosswise multiplication.

Calculation and check of the answer

This method comprises equations with rational numbers, and is intended for students in order to calculate and check the answer of an equation before it is solved by a formula. The left side of the equation is factorised and we get two binomials, that are converted into answers. When the binomial has a plus sign it has to be changed into minus in the answer, e.g. $3x + 2$, i.e. $x = -2/3$. Check first if the middle coefficient is an even or an odd number, since it tells us how the coefficients are to be split up into factors. Make also sure that the result after crosswise multiplication is identical with the polynomial, and that the algorithm is followed in the example as shown below.

1. $5x^2 + 18x - 8 = 0$
   $5x - 2$
   $x + 4 = 5x^2 + 18x - 8$ $\qquad\qquad\qquad$ $x_1 = 2/5, \quad x_2 = -4$

The left side of the equation is factorised and $5x^2$ is split into $5x/x$ and 8 into 2/4 as 18 is an even number. Put the factors $5x/x$ under each other and provide the factors 2/4 with a plus or minus sign, so the result after cross multiplication is the same as the polynomial.

2. $12x^2 + 33x + 18 = 0$
   $3x + 6$
   $4x + 3 = 12x^2 + 33x + 18$ $\qquad\qquad$ $x_1 = -2, \quad x_2 = -3/4$

The example shows that it is just as simple when the coefficients are large. We note that two odd factors are multiplied crosswise, since the middle coefficient is an odd number.

3. $3x^2 + 5x - 8 = 0$
   $3x + 8$
   $x - 1 = 3x^2 + 5x - 8$ $\qquad\qquad\qquad$ $x_1 = -8/3, \quad x_2 = 1$

When the middle coefficient is an odd number 8 must be split into 1/8.

4. $x^2 - 8x + 7 = 0$ $\qquad\qquad\qquad\qquad$ $(-1x - 7x = -8x)^*$
   $x - 1$
   $x - 7 = x^2 - 8x + 7$ $\qquad\qquad\qquad$ $x_1 = 1, \quad x_2 = 7$

*If the first coefficient is 1 the last term has to be split up so that the sum of the factors is equal to the coefficient of the x–term. The product 7 is the same as the last term.

5. $x^2 + 6x - 12 = 0$ $\qquad\qquad$ $x_1 = -3 + \sqrt{21}, \quad x_2 = -3 - \sqrt{21}$
   $x - 1,58$ $\qquad\qquad\qquad\qquad$ or $x_1 = 1,58, \quad x_2 = -7,58$
   $x + 7,58 = x^2 + 6x - 12$

Since the equation contains irrational numbers, it has to be solved by a formula, and this can easily be done mentally by the new formula. $x = -3 \pm\sqrt{9 + 1\cdot 12} = -3 \pm\sqrt{21}$
NB! Use a formula as soon as you think the equation contains irrational numbers.

6. $2x^2 + 17x + 8 = 0$

We solve this equation by mental calculation. The left part of the equation is factorised and $2x^2$ is split up into $2x/x$ and 8 is split up into 1/8, since 17 is an odd number.

$2x + 1$

$x + 8 = 2x^2 + 17x + 8$ $\qquad\qquad\qquad\qquad x_1 = -1/2,\ x_2 = -8$

Visualize the position of the factors carefully and let it stick to the retina, and when the binomials are positive x will be negative.

7. $+17x +$ is just known in the equation.

We will try to get two binomials and when these are multiplied vertically and crosswise, we will get a polynomial where 17 is included. We multiply two factors, the product of which is less but not larger than the middle coefficient. The first coefficient ahead of x^2 we choose to be 2 and 8 in the last term. We split $2x^2$ into $2x/x$ and 8 is split into 1/8 and by crosswise multiplication we get:

$2x + 1$

$x + 8 = 2x^2 + 17x + 8$ $\qquad\qquad\qquad\qquad x_2 = -8 \quad x_1 = -1/2$

Note! This example shows that the middle coefficient contains all information about the origin of the equation. It can be compared to a cell nucleus containing the whole genetic material, i.e. its DNA.

All information about the origin of an equation is found in the next highest coefficient, so that means that a third degree equation contains all information about its origin in the next highest coefficient x^2.

Summary

We have seen that the middle coefficient has shown us how to split up the coefficients into factors. We have seen that the answer to an equation can be both calculated and checked before it is solved by a formula. We have seen an example of how to solve an equation mentally and that only the middle coefficient is needed to solve an equation. This is due to the two binomials, which are the origin of the equation, can be traced up by the split up rule. This rule can be compared to a key that unlocks a safe where the origin of the equation is hidden.

By factorising and the new formula it will be easy and fast to solve quadratic equations. As the equations are mostly solved mentally, it will stimulate both your flexibility and creativity, and it will also increase your self-confidence, satisfaction and joy. As well as short term, medium term and long term memory are all activated by mental calculation.

Quadratic equations are solved by factorising

An equation describes that two mathematical expressions on both sides of the equal sign are equal $(6 + 4 = 10)$. We solve the following equation.

$x^2 - 4x - 12 = 0$

The left hand side of the equation is factorised and x^2 is split into x/x and 12 into 2/6, since the middle coefficient is an even number. If the middle coefficient is to be $- 4$, we have to multiply $x (- 6) + x \cdot 2$ crosswise and the factors $(x + 2)(x - 6)$ are as shown.

$x + 2$

$x - 6 = x^2 - 4x - 12$

The left hand side LHS= $(x + 2)(x - 6)$ and the right hand side RHS= 0.

$(x + 2)(x - 6) = 0$.

The only way for a product to be zero is that one of the factors is zero. Since we know that the product is zero in this example, either $(x + 2)$ is zero or $(x - 6)$ is zero.

We can use this in our equation $(x + 2)(x - 6) = 0$.
The first factor $= 0$ gives $x + 2 = 0$, i.e. $x_1 = - 2$.
The second factor $= 0$ gives $x - 6 = 0$, i.e. $x_2 = 6$.
The quadratic equation has two roots $x_1 = - 2$, $x_2 = 6$.
The roots can be checked by a test.

We put $x_1 = - 2$ in the equation $(x + 2)(x - 6) = 0$ gives the left side:
LHS $= (- 2 + 2)(-2 - 6) = - 8 \cdot 0 = 0$.
The right hand side RHS $= 0$.

We put $x_2 = 6$ in the equation $(x + 2)(x - 6) = 0$, gives the left hand side:
LHS $= (6 + 2)(6 - 6) = 8 \cdot 0 = 0$.
The right hand side RHS $= 0$.

We can see that the value of the equation on the left hand side is equal to the value on the right hand side. The solutions $x = - 2$ and $x = 6$ satisfies the equation.

$x_1 = - 2$, $x_2 = 6$.

Example 1 $3x^2 + 13x + 12 = 0$

The left side of the equation is factorised and $3x^2$ is split into 3x/x and 12 into 3/4. As the middle coefficient is 13 an odd number in the first term is crosswise-multiplied by an odd factor in the last term. If the middle coefficient shall be 13, 3x has to be multiplied crosswise by 3, i.e. $3x \cdot 3 + x \cdot 4$. The factors are $(3x+4)(x+3)$ as shown below.

3x + 4

 x + 3 = $3x^2 + 13x + 12$

The left side = $(3x+4)(x+3)$ and the right side = 0
$(3x+4)(x+3) = 0$.
The first factor = 0 gives $3x+4 = 0$, i.e. $x_1 = -4/3$
The second factor = 0 gives $x+3 = 0$, i.e. $x_2 = -3$
$x_1 = -4/3$, $x_2 = -3$

Example 2 $2x^2 - 16x + 14 = 0$

The left side of the equation is factorised and $2x^2$ is split into 2x/x and 14 into 2/7. Since the middle coefficient is negative and the constant term positive, both 2 and 7 must be prefixed by a minus sign. Multiply 2x by-7 as $2x \cdot 7$ is nearer to $-16x$ than to $2x \cdot 2$. Multiply crosswise $2x(-7) + x(-2) = -16x$ and the factors are $(2x-2)(x-7)$ as shown.

2x – 2

x – 7 = $2x^2 - 16x + 14$

 The left side = $(2x - 2)(x - 7)$ and the right side = 0
$(x - 1)(x - 7) = 0$.
The first factor = 0 gives $2x - 2 = 0$, i.e. $x_1 = 1$
The second factor = 0 gives $x - 7 = 0$, i.e. $x_2 = 7$
$x_1 = 1$, $x_2 = 7$

Example 3 $6x^2 + 17x + 12 = 0$

The left side of the equation is factorised and $6x^2$ is split into 3x/2x and 12 into 3/4 as 17 is an odd number. Multiply the two odd factors 3x and 3 crosswise as shown below.

$6x^2 + 17x + 12$

$3x + 4$

$2x + 3 = 6x^2 + 17x + 12$

The left side $= (3x + 4)(2x + 3)$ and the right side $= 0$.

$(3x + 4)(2x + 3) = 0$.

The first factor $= 0$ gives $3x+4 = 0$, i.e. $x_1 = -4/3$.

The second factor $= 0$ gives $2x+3 = 0$, i.e. $x_2 = -3/2$.

$x_1 = -4/3$ and $x_2 = -3/2$.

Quadratic equations are solved mentally

Quadratic equations can easily be solved by the dividing rule and the factor rule. Note if the middle coefficient is an odd or an even number.

Example 1 $2x^2 - 16x + 14 = 0$

Multiply 2x by 7. Visualize the location of the factors and keep it in your mind.
$x_1 = 7, x_2 = 1$
$2x - 2$
$x \ - 7$

By mental calculation it is easier to use the factor rule even if the middle coefficient is an odd number. After each example the equation is solved by factorising.

$2x^2 - 16x + 14 = 0$

The left side of the equation is factorised and $2x^2$ is split into 2x/x and 14 into 2/7, since the middle coefficient is an even number. Cross-multiply 2x by 7 and x by 2 and we get the factors $(x - 7)(2x - 2)$ as shown below.
$2x^2 - 16x + 14 = 0$
$2x - 2$
$x \ - 7 = 2x^2 - 16x + 14$

The left side $= (2x-2)(x-7)$ and the right side $= 0$
$(2x - 2)(x - 7) = 0$
The first factor $= 0$ gives $2x - 2 = 0$, i.e. $x_1 = 1$
The second factor $= 0$ gives $x - 7 = 0$, i.e. $x_2 = 7$

$x_1 = 1, \quad x_2 = 7$

Example 3 $2x^2 + 17x + 8 = 0$

Cross multiply 2x by 8. Visualize carefully as shown and keep it in your mind.
$x_1 = -1/2$, $x_2 = -8$

$2x^2 + 17x + 8 = 0$
2x +1
 x + 8

$2x^2 + 17x + 8 = 0$

The left side of the equation is factorised and $2x^2$ is split into 2x/x and 8 into 8/1, since the middle coefficient is an odd number. Two odd factors x and 1 are cross-multiplied and we get the factors $(2x +1)(x + 8)$.
$2x^2 + 17x + 8 = 0$
2x + 1
 x + 8 $= 2x^2 + 17x + 8$

The left side $= (2x + 1)(x + 8)$ and the right side $= 0$
$(2x + 1)(x + 8) = 0$
The first factor $= 0$ gives $2x + 1 = 0$, i.e. $x_1 = -1/2$
The second factor $= 0$ gives $x + 8 = 0$, i.e. $x_2 = -8$
$x_1 = -1/2$, $x_2 = -8$

<u>NB!</u> In this equation $2x^2 + 17x + 8 = 0$, the product of the numerators in the roots is the same as the constant term 8, and the product of the denominators in the roots is the same as the first coefficient 2. Through cross-multiplying we get the middle coefficient 17x.
2x +1
 x + 8 $= 2x^2 + 17x + 8$ $x_1 = -1/2$, $x_2 = -8$

So when we multiply crosswise the numerator and the denominator we get the middle coefficient 17 or $\dfrac{1 \cdot 8}{2 \cdot 1}$

Example 4 $2x^2 + 7x + 5 = 0$

Mentally we multiply 2x crosswise by 1 and x by 5. $x_1 = -1, x_2 = -5/2$

$2x^2 + 7x + 5 = 0$
$2x + 5$
$x + 1$

The left side of the equation is factorised and 2x is split into 2x/x and 5 is split into 5/1. So as not to exceed the middle coefficient 7, we multiply $2x \cdot 1 + x \cdot 5 = 7x$, and we get the factors $(2x + 5)(x + 1)$ as shown below.
$2x^2 + 7x + 5 = 0$
$2x + 5$
$x + 1 = 2x^2 + 7x + 5$

The left side $= (2x + 5)(x + 1)$ and the right side $= 0$
$(2x + 5)(x + 1) = 0$.
The first factor $= 0$ gives $2x + 5 = 0$, i.e. $= -5/2$
The second factor $= 0$ gives $x + 1 = 0$, i.e. $x_2 = -1$
$x_1 = -5/2, x_2 = -1$

Example 4 $3x^2 + 14x + 8 = 0$

Mentally 3x is cross-multiplied by 4 and x by 2. $x_1 = -2/3, x_2 = -4$
$3x^2 + 14x + 8$
$3x + 2$
$x + 4$

$3x^2 + 14x + 8 = 0$

The left side of the equation is factorised and $3x^2$ is split into 3x/x and 8 into 2/4, since 14x is an even number. Cross-multiply 3x by 4 and x by 2. As shown below the factors are $(3x + 2)(x + 4)$.
$3x^2 + 14x + 8$
$3x + 2$
$x + 4 = 3x^2 + 14x + 8$

The left side $= (3x + 2)(x + 4)$ and the right side $= 0$
$(3x + 2)(x + 4) = 0$.
The first factor $= 0$ gives $3x + 2 = 0$, i.e. $x_1 = -2/3$
The second factor $= 0$ gives $x + 4 = 0$, i.e. $x_2 = -4$
$x_1 = -2/3, x_2 = -4$

Example 6 $6x^2 + 13x + 6 = 0$

Mentally we multiply 3x crosswise by 3 and $x_1 = -2/3$, $x_2 = -3/2$
$6x^2 + 13x + 6 = 0$
$3x + 2$
$2x + 3 = 6x^2 + 13x + 6$

$6x^2 + 13x + 6 = 0$

The left side of the equation is factorised and $6x^2$ split into 3x/2x and 6 is split into 3/2. Since the middle coefficient is an odd number one odd number in the first term is multiplied by an odd number in the last term. When the odd factors are multiplied crosswise we get $3x \cdot 3 + 2x \cdot 2 = 13x$, and the factors $(3x + 2)(2x + 3)$ are verified as shown.

$6x^2 + 13x + 6 = 0$
$3x + 2$
$2x + 3 = 6x^2 + 13x + 6$

The left side $= (3x + 2)(2x + 3)$ and the right side $= 0$
$(3x + 2)(2x + 3) = 0$
The first factor $= 0$ gives $3x + 2 = 0$, i.e. $x_1 = -2/3$
The second factor $= 0$ gives $2x + 3 = 0$ i.e. $x_2 = -3/2$
$x_1 = -2/3$, $x_2 = -3/2$

An example of a polynomial containing plus and minus

Let us factorize a polynomial containing both plus and minus signs, $8x^2 + 2x - 15$. We want to find two binomials and when these are cross-multiplied together, we should get this polynomial $8x^2 + 2x - 15$.

The first coefficient $8x^2$ is split into 4x/2x, as the middle coefficient is an even number and 15 into 3/5.

In accordance with the factor rule the largest factor in the first term 4x is multiplied. We shall choose if 4x is to be multiplied by 3 or by 5 and if it should be a minus or a plus sign. Since the middle coefficient 2 is positive, we will choose between +3 and + 5. Is it $4x \cdot 3 = 12x$ or $4x \cdot 5 = 20x$ that are nearest 2x? We can see that $4x \cdot 3$ is nearest to 2x. It means that $4x \cdot 3$ is multiplied crosswise by $2x\ (-5)$, which is numerically less because of the choice we have made. We have confirmed that the first binomial (4x–5) and the second one is (2x + 3). To verify we multiply crosswise and we get the factors (4x–5) and (2x + 3) as shown below.

$8x^2 + \underline{2}x - 15$
4x – 5
$2x + 3 = 8x^2 + \underline{2}x - 15$

Note how the middle coefficient 2 is found here as well.
$x_1 = 5/4, \quad x_2 = -3/2$

Another way of solving the equation is:
$8x^2 + 2x - 15 = 0$
The left side of the equation is factorised and $8x^2$ is split into 4x/2x and 15 into 3/5. Put 4x and 2x under each other, and place and supply the factors 3 and 5 with plus or minus so that the result after crosswise multiplication is the same as the polynomial as shown.

$8x^2 + 2x - 15$
4x – 5
$2x + 3 = 8x^2 + 2x - 15$
$x_1 = 5/4, \quad x_2 = -3/2$

The core of a polynomial

We have seen that the middle coefficient is the sum of two products and how important the middle coefficient is when we factorise. It gives us a signal how the coefficients are to be split and it is also the origin of all formulas and rules. The middle coefficient plays then a central role and is the core of a polynomial. It can be compared to a cell nucleus containing the entire genetic material, i.e. its DNA.

To emphasize how important the middle coefficient is, a few examples will show how it is possible to factorise when only the middle coefficient is known, provided with plus or minus. The middle coefficient can of course be the same in several polynomials, but given that the coefficients are integers they can also be unambiguous.

Example 1 The outer coefficients are unknown

$+ 17x +$

When the middle coefficient is an odd number, two odd factors have been multiplied crosswise. We will now try to find these two binomials and when these are multiplied together vertically and crosswise, we shall get a polynomial where 17x is included. We will multiply two factors, which are lower but not larger than 17x. The first coefficient ahead of x^2 we choose 2 and 8 in the last term. We split $2x^2$ into $2x/x$ and 8 into 1/8. By crosswise multiplication we get as follows.

2x +1

$x + 8 = 2x^2 + 17x + 8$

When we convert the binomials 2x+1 and x + 8, we get: $x_1 = -\frac{1}{2}$, $x_2 = -8$

We check the answer and solve the equation

$2x^2 + 17x + 8 = 0$

$2x = -8.5 \pm \sqrt{72{,}25 + 2(-8)} = -8.5 \pm \sqrt{56{,}25} = -8.5 \pm 7.5$

$x_1 = -\frac{1}{2}$, $x_2 = -8$.

The example clearly shows that the middle coefficient is the core of the polynomial.

Example 2 The first coefficient is unknown

+19x + 5
We split 5 into 5 and 1 as shown.
+19x + 5
     5
     1
When the middle coefficient is an odd number, 5 or 1 has to be multiplied by one factor in the first term. In order not to exceed 19x, we choose to cross-multiply 3x by 5 and we get the factors $3x \cdot 5 + 4x \cdot 1 = 19x$ as shown below.
3x + 1
$4x + 5 = 12x^2 + 19x + 5$

When we convert the binomials 3x+1 and 4x + 5, we get: $x_1 = - 1/3$ $x_2 = - 5/4$.

We check the answer and solve the equation, $12x^2 + 19x + 5 = 0$

$$12x = - 9.5 \pm \sqrt{90.25 + 12(-5)} = - 9.5 \pm \sqrt{30.25} = - 9.5 \pm 5.5.$$

$x_1 = - 1/3, x_2 = - 5/4.$

Example 3 The constant term is unknown

$7x^2 + 10x +$
We split $7x^2$ into 7x and x.
$7x^2 + 10x +$
7x +
 x +
The middle coefficient may not exceed 10, and 7x has to be multiplied by 1. By cross-wise multiplication we get $7 \cdot 1 + x \cdot 3 = 10x$ as shown.
7x + 3
 x + 1 = $7x^2 + 10x + 3$

When we convert the binomials 7x + 3 and x + 1, we get: $x_1 = - 3/7, \; x_2 = -1$

We check the answer and solve the equation $7x^2 + 10x + 3 = 0$

$$7x = -5 \pm \sqrt{25 + 7(-3)} = - 5 \pm \sqrt{4} = - 5 \pm 2$$

$x_1 = - 3/7, x_2 = -1$

In a cubic polynomial the core of the polynomial is found in the coefficient ahead of x^2, in a fourth degree polynomial ahead of x^3 and in a fifth degree polynomial ahead of x^4.

Example 4 The outer coefficients are known in a cubic equation.

$x^3 + 20$

When prime factorising of 20 we get: $20 = 2 \cdot 10 = 2 \cdot 2 \cdot 5$. When the factors are added $2+2+5$, we get 9 corresponding $9x^2$. When these three factors are multiplied together we get. $x^3 + 9x^2 + 24x + 20$ and $x_1 = x_2 = -2$, $x_3 = -5$, see page 59.

Example 5 The outer coefficients are known in a fifth degree equation.

$x^5 + 36$

Prime factorising of 36 is: $36 = 2 \cdot 18 = 2 \cdot 2 \cdot 9 = 2 \cdot 2 \cdot 3 \cdot 3$, but since we have five factors and the product has to be 36, the fifth factor is 1. When we add the factors $1+2+2+3+3$, we get 11 corresponding $11x^4$. When multiplying the factors we get: $x^5 + 11x^4 + 47x^3 + 97x^2 + 96x + 36$. See p. 86.
$x_1 = -1$, $x_2 = x_3 = -2$, $x_4 = x_5 = -3$.

Example 6

$5x^5 + 168x + 36$

The factors are the same as example 5, but we have to use the factor rule from chapter 3 since the first coefficient is larger than 1. The factor rule says:

" Multiply the highest factor in the first term by the factors in the constant term", i.e.
$5x \cdot 1 \cdot 2 \cdot 2 \cdot 3 \cdot 3 = 180x > 168x$.

" If the product is larger than the coefficient in the variable term x, the highest factor in the first term shall be part of the binomial with the next smallest factor in the last term". It means the first binomial is $5x + 2$. When the factors are multiplied we get as follows.

$5x^5 + 67x^4 + 163x^3 + 254x^2 + 168x + 36$, see page 86.

$x_1 = -2/5$, $x_2 = -1$, $x_3 = -2$, $x_4 = x_5 = -3$

When one of the outer coefficients is 1

It is easy to factorise when one of the outer coefficients is 1, as the middle coefficient is the sum of two products. The other coefficient can then be split into such a way that the sum of the factors is the same as the middle coefficient as shown below.

$6x^2 + 5x + 1 = 0$

The left side of the equation is factorised and $6x^2$ is split into $3x/2x$ and 1 into $1/1$. Since the factors in the last term are just 1, you need not to multiply crosswise, as the middle coefficient 5 is the sum of two products $3x \cdot 1 + 2x \cdot 1 = 5x$ and we get the factors $(3x+1)$ and $(2x+1)$.

$3x + 1$
$2x + 1 = 6x^2 + 5x + 1$
$x_1 = -1/3$, $x_2 = -1/2$.

The same ratio can be applied to a third-, fourth- and fifth degree equation when one of the outer coefficients is 1 and in a similar way the coefficients of a third degree equation are split into such a way that the sum of the three factors are the same as the coefficient of the variable term x^2, corresponding to x^3 in a forth degree equation and to x^4 in a fifth degree equation.

Identical binomials

A coefficient can be split into different ways, and gives the same answer. The following examples show different alternatives as shown below.

Example 1
$4x^2 + 12x + 8 = 0$

The left side of the equation is factorised and $4x^2$ is split into $4x/x$ and 8 into $2/4$. Multiply $4x$ by 2 as $4x \cdot 2$ is less than $12x$. Cross-multiply $4x \cdot 2 + x \cdot 4 = 12x$. The factors are $(4x + 4)(x + 2)$.

Alternative 1
$4x + 4$
$\ \ x + 2 = 4x^2 + 12x + 8 = (4x + 4)(x + 2) = 4(x + 1)(x + 2)$.
$x_1 = -1$, $x_2 = -2$

Alternative 2

$4x^2 + 12x + 8 = 0$
The left side of the equation is factorised and $4x^2$ is split into 2x/2x and 8 into 2/4. Multiply 2x by 4 since 2x · 4 is less and nearer to 12x than 2x · 2. Cross-multiply 2x · 4 + 2x · 2 = 12x and the factors are (2x +2)(2x +4) as shown.
2x + 2
$2x + 4 = 4x^2 + 12x + 8 = (2x + 2)(2x + 4) = 2(x + 1)\ 2\ (x + 2) = 4\ (x + 1)(x + 2)$
$x_1 = -1\ x_2 = -2$

Alternative 3
$4x^2 + 12x + 8 = 0$
The left side of the equation is factorised and $4x^2$ is split into 2x/2x and 8 into 8/1. Multiply 4x by 1, as 4x · 1 is less and nearer to 12x than 4x · 8. Cross-multiply 4x · 1 + x · 8 = 12x and the factors are (x + 1)(4x+8).
4x + 8
$\ \ x + 1 = 4x^2 + 12x + 8 = (x + 1)(4x + 8) = (x + 1) · 4\ (x + 2) = 4\ (x + 1)(x + 2).$
$x_1 = -1,\ x_2 = -2$

Example 2
$12x^2 + 33x + 18 = 0$
The left side of the equation is factorised and $12x^2$ is split into 12x/x, 4x/3x and 18 into 3/6 and 2/9. As a change we combine and multiply 4x · 6 + 3x · 3 = 33x and 12x · 2 + x · 9 = 33x. Cross-multiply 4x · 6 + 3x · 3 = 33x and 12x · 2 + x · 9 = 33x.

Alternative 1
$12x^2 + 33x + 18 = 0$
4x + 3
$3x + 6 = 12x^2 + 33x + 18 = (4x + 3)(3x + 6) = (4x + 3)\ 3\ (x + 2) = 3(4x + 3)(x + 2).$
$x_1 = -3/4\ ,\ x_2 = -2$

Alternative 2

Alt. 2

$12x^2 + 33x + 18 = 0$

$12x + 9$

$\quad x + 2 = 12x^2 + 33x + 18 = (12x + 9)(x + 2) = 3(4x + 3)(x + 2).$

$x_1 = -3/4$, $x_2 = -2$

Example 3

Alternative 1

$6x^2 + 24x + 18 = 0$

The left side of the equation is factorised and $6x^2$ is split into 3x/2x, 6x/x and 18 into 3/6 and 2/9. Combine and multiply $3x \cdot 6 + 2x \cdot 3 = 24x$, $6x \cdot 3 + x \cdot 6 = 24x$ and $2x \cdot 9 + 3x \cdot 2 = 24x$ and cross-multiply as shown.

$6x^2 + 24x + 18$

$3x + 3$

$2x + 6 = 6x^2 + 24x + 18 = (3x + 3)(2x + 6) = 3(x + 1) 2 (x + 3) = 6(x + 1)(x + 3)$

$x_1 = -1$, $x_2 = -3$

Alternative 2

$6x^2 + 24x + 18 = 0$

$6x + 6$

$\quad x + 3 = 6x^2 + 24x + 18 = (6x + 6)(x + 3) = 6 (x + 1)(x + 3).$

$x_1 = -1$, $x_2 = -3$

Alternative 3

$6x^2 + 24x + 18 = 0$

$2x + 2$

$3x + 9 = 6x^2 + 24x + 18 = (2x + 2)(3x + 9) = 2 (x + 1) 3 (x + 3) = 6 (x + 1)(x + 3)$

$x_1 = -1$, $x_2 = -3$

Completing the square

Instead of using a formula some quadratic equations can be solved through completing a small square. That means that a quadratic polynomial is written in quadratic form. We simply add a "square" on both sides of the equal sign in an equation. We can factorise the left side of the equation by using the formula $(a + b)^2 = a^2 + 2ab + b^2$ and then solve the equation. The coefficient ahead of x^2 has to be 1.

Example 1

$x^2 + 6x - 16 = 0$ We add 16 on both sides of the equal sign.
$x^2 + 6x - 16 + 16 = 0 + 16$
$x^2 + 6x = 16$

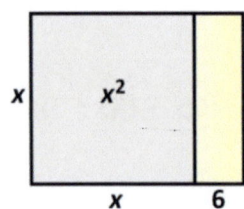

In order to clarify what completing the square means, we will make a geometrical interpretation of the quadratic equation. In the figure we have a rectangle which consists of two parts, one square x^2 where length and width are x and a rectangle, where the width is 6 and the length is x, and we get the total area to be $x^2 + 6x = 16$.

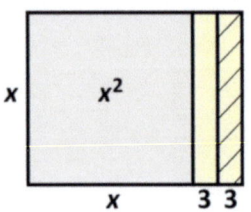

We halve the little rectangle in such a way so that the area in each part is $3 \cdot x$. We put one part on the top of the square as shown in the next figure.

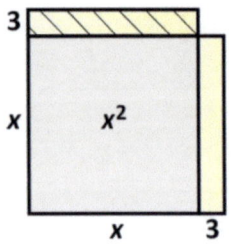

We have got a new figure, which is almost a square, but with the same area as before, i.e.16.

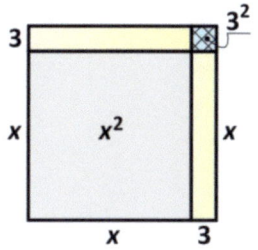

To get a square, we add a small square and that is why it is called completing the square. We complete a small area 3^2, which is added on both sides of the equal sign, and the area of the new square is x^2 and plus $6 \cdot x$ i.e. the two rectangles $(3 \cdot x + 3 \cdot x)$ plus the small square 3^2.

The total area of the new square is now $x^2 + 6x + 3^2 = 16 + 3^2 = 25$. The length of the new square is $x + 3$ and since it is a square the width is also $x + 3$.

We multiply $x + 3$ by $x + 3$ and get $(x + 3)^2 = x^2 + 6x + 9$. Using the formula $(a + b)^2$ we get: $= a^2 + 2ab + b^2$, and the left side $x^2 + 6x + 9$ is factorised as shown below.

$(x + 3)^2 = 25$

$x + 3 = \pm\sqrt{25}$

$x + 3 = \pm 5$

$x_1 = -8, x_2 = 2$

Example 2

$x^2 + 4x - 5 = 0$.   We add 5 on both sides of the equal sign and get:

$x^2 + 4x - 5 + 5 = 0 + 5$

$x^2 + 4x = 5$

On both sides of the equal sign we complete with a small square 2^2 and we get:

$x^2 + 4x + 2^2 = 5 + 2^2$

$x^2 + 4x + 4 = 9 = (x + 2)^2$

By using the formula $(a + b)^2 = a^2 + 2ab + b^2$, we factorise the left side as shown.

$(x + 2)^2 = 9$

$x + 2 = \pm\sqrt{9}$

$x + 2 = \pm 3$

$x_1 = -5, x_2 = 1$.

Example 3

$x^2 + 4x - 3 = 0$. We add 3 on both sides of the equal sign and get:
$x^2 + 4x - 3 + 3 = 0 + 3$
$x^2 + 4x = 3$

We complete both sides with a small square 2^2 and we get:
$x^2 + 4x + 2^2 = 3 + 2^2$
$x^2 + 4x + 4 = 7$
$x^2 + 4x + 4 = (x + 2)^2$
By using the formula $(a + b)^2 = a^2 + 2ab + b^2$, the left side is factorised as shown.
$(x + 2)^2 = 7$
$x + 2 = \pm\sqrt{7}$
$x_1 = -2 + \sqrt{7}$, $x_2 = -2 - \sqrt{7}$

Alternative methods

Example 1

$x^2 + 6x - 16 = 0$

$x = 3 \pm\sqrt{9 + 1 \cdot 16} = 3 \pm\sqrt{25} = 3 \pm 5$ $x_1 = -8, x_2 = 2$ or

$x + 8$
$x - 2 = x^2 + 6x - 16$ $x_1 = -8, x_2 = 2$ $x_1 = -8, x_2 = 2$

Example 2

$x^2 + 4x - 5 = 0$

$x = -2 \pm\sqrt{4 + 1 \cdot 5} = -2 \pm\sqrt{9} = 2 \pm 3$ $x_1 = -5, x_2 = 1$ or

$x + 5$
$x - 1 = x^2 + 4x - 5$ $x_1 = -5, x_2 = 1$

Example 3

$x^2 + 4x - 3 = 0$.

$x = -2 \pm\sqrt{4 + 1 \cdot 3} = -2 \pm\sqrt{7}$ $x_1 = -2 + \sqrt{7}$, $x_2 = -2 - \sqrt{7}$

Exercises with answers

Example 1 $3x^2 + 25x + 8 = 0$

$3x +1$ $x_1 = -1/3 , x_2 = -8$
$x + 8 = 3x^2 + 25x + 8$

The left side of the equation is factorised and $3x^2$ is split into $3x/x$ and 8 into $1/8$, since the middle coefficient 25 is an odd number. Put the factors under each other and provide the factors $1/8$ so the result after crosswise multiplication is the same as the polynomial. As the middle coefficient is an odd number, the two odd numbers x and 8 have to be cross-multiplied.

The Vedic method is based on proportionality

$3x^2 + 25x + 8$
Split up the middle coefficient 25 into two parts, 24 and 1, giving two equal ratios. This is combined as shown with the other coefficients. The ratio 3 : 1 gives one factor $(3x+1)$ and the second one is using the same parts of 25 in reverse order as shown. Thus $(x + 8)$ is the second factor.

$3x^2 + 25x + 8$
$3 : 1 = 24 : 8 = (3x +1)$
$3x^2 + 25x + 8$
$3 : 24 = 1 : 8 = (x + 8)$
$\therefore\ 3x^2 + 25x + 8 = (3x + 1)(x + 8).$

Ratio and proportion are fundamental to mathematics. Calculations can often be simplified by means of the proportionality. This extends the variety of methods we have for multiplication and division, and adds to the enjoyment of mental calculation.

Example a) $440 \div 5 = 880 \div 10 = 88$
Doubling is easier than dividing by 5.
Example b) $35 \times 42 = 70 \times 21 = 1470$
Doubling and halving leads to the multiplication of two smaller numbers.

Example 2 $2x^2 + 10x + 8 = 0$

$2x + 2$ $x_1 = -1$, $x_2 = -4$
$x + 4 = 2x^2 + 10x + 8$

The left side of the equation is factorised and $2x^2$ is split up into 2x/x and 8 into 2/4 since 10x is an even number. Put the factors under each under and provide the factors 2/4 so the result after cross-multiplication is the same as the polynomial.

Example 3 $12x^2 + 17x + 6 = 0$

$4x + 3$ $x_1 = -3/4$, $x_2 = -2/3$
$3x + 2 = 12x^2 + 17x + 6$

The left side of the equation is factorised and $12x^2$ is split up into 4x/3x and 6 into 2/3 since the middle coefficient is an odd number. Put the factors under each under and provide the factors so that the result after crosswise multiplication is the same as the polynomial.

Example 4 $5x^2 + 22x + 8 = 0$

$5x + 2$ $x_1 = -2/5$, $x_2 = -4$
$x + 4 = 5x^2 + 22x + 8$

The left side of the equation is factorised and $5x^2$ is split up into 5x/x and 8 into 2/4 since 22x is an even number. Put the factors under each under and provide the factors so that the result after crosswise multiplication is the same as the polynomial.

Example 5 $5x^2 + 21x + 18 = 0$

$5x + 6$ $x_1 = -6/5$, $x_2 = -3$
$x + 3 = 5x^2 + 21x + 18$

The left side of the equation is factorised and $5x^2$ is split up into 5x/x and 18 into 3/6 since 21x is an odd number. Put the factors under each other and provide the factors 3/6 so the result after crosswise multiplication is the same as the polynomial

Example 6 $5x^2 - x - 18 = 0$

$5x + 9$ $x_1 = -9/5, \ x_2 = 2$
$\ \ x - 2 = 5x^2 - x - 18$

The left side of the equation is factorised and $5x^2$ is split into 5x/x and 18 into 2/9 since x is an odd number. Put the factors under each other and provide the factors 2/9 so the result after crosswise multiplication is the same as the polynomial.

Example 7 $4x^2 + 12x + 8 = 0$

$4x + 4$ $x_1 = -1, \ x_2 = -2$
$\ \ x + 2 = 4x^2 + 12x + 8$

The left side of the equation is factorised and $4x^2$ is split into 4x/x and 8 into 2/4. Put the factors under each other and provide the factors 4/2 so the result after crosswise multiplication is the same as the polynomial

Example 8 $4x^2 - 14x + 6 = 0$

$4x - 2$ $x_1 = 1/2, x_2 = 3$
$\ \ x - 3 = 4x^2 - 14x + 6$

The left side of the equation is factorised and $4x^2$ is split into 4x/x and 6 to 2/3. Put the factors under each other and provide the factors 2/3 so the result after crosswise multiplication is the same as the polynomial.

Example 9 $7x^2 + 18x + 8 = 0$
$7x + 4$ $x_1 = -4/7, x_2 = -2$
$\ \ x + 2 = 7x^2 + 18x + 8$

The left side of the equation is factorised and $7x^2$ is split into 7x/x and 8 into 2/4. Put the factors under each other and provide the factors 4/2 so the result after crosswise multiplication is the same as the polynomial.

Example 10 $x^2 + 13x - 48 = 0$

    x +16 $x_1 = -16$, $x_2 = 3$

    x − 3 = $x^2 + 13x - 48$

The left side of the equation is factorised and x^2 is split into x/x and 48 into 3/16 since the middle coefficient 13 is an odd number. Put the factors under each other and provide the factors 3/16 so the result after crosswise multiplication is the same as the polynomial.

It is easy to factorise when one of the outer coefficients is 1, since the middle coefficient is the sum of two products. The other coefficient can then be split into such a way that the sum of the factors is the same as the middle coefficient as x(−3) + x ·16 = 13x.

Example 11 $6x^2 + 5x + 1 = 0$ When one outer coefficient is 1 se example 10.

    2x + 1 $x_1 = -1/2$, $x_2 = -1/3$

    3x + 1 = $6x^2 + 5x + 1$

The left side of the equation is factorised and $6x^2$ is split into 3x/2x and 1 into 1/1. Since the last coefficient is 1 it is easy.

Example 12 $6x^2 + 20x + 14 = 0$

    3x + 7 $x_1 = -7/3$, $x_2 = -1$

    2x + 2 = $6x^2 + 20x + 14$

The left side of the equation is factorised and $6x^2$ is split into 3x/2x and 14 into 2/7. Put the factors under each other and provide the factors 2/7 so the result after crosswise multiplication is the same as the polynomial.

Example 13 $10x^2 - 16x + 6 = 0$

    5x − 3 $x_1 = 3/5$, $x_2 = 1$

    2x − 2 = $10x^2 - 16x + 6$

The left side of the equation is factorised and $10x^2$ is split into 5x/2x and 6 into 2/3. Put the factors under each other and provide the factors 3/2 so the result after crosswise multiplication is the same as the polynomial.

Example 14 $4x^2 + 14x + 6 = 0$

$4x + 2$ $x_1 = -1/2, x_2 = -3$

$x + 3 = 4x^2 + 14x + 6$

The left side of the equation is factorised and $4x^2$ is split into $4x/x$ and 6 into 2/3. Put the factors under each other and provide the factors 2/3 so the result after crosswise multiplication is the same as the polynomial.

Example 15 $6x^2 + 19x + 14 = 0$

$6x + 7$ $x_1 = -7/6, x_2 = -2$

$x + 2 = 6x^2 + 19x + 14$

The left side of the equation is factorised and $6x^2$ is split into $6x/x$ and 14 into 2/7. Put the factors under each other and provide the factors 7/2 so the result after crosswise multiplication is the same as the polynomial.

Example 16 $2x^2 - 16x + 14 = 0$

$2x - 2$ $x_1 = 1, x_2 = 7$

$x \ - 7 = 2x^2 - 16x + 14$

The left side of the equation is factorised and $2x^2$ is split into $2x/x$ and 14 into 2/7. Put the factors under each other and provide the factors 2/7 so the result after crosswise multiplication is the same as the polynomial.

Example 17 $6x^2 + 7x - 3 = 0$

$3x - 1$ $x_1 = 1/3 , x_2 = -3/2$

$2x + 3 = 6x^2 + 7x - 3$

The left side of the equation is factorised and $6x^2$ is split into $3x/2x$ and 3 into 3/1. Put the factors under each other and provide the factors 2/7 so the result after crosswise multiplication is the same as the polynomial.

Example 18 $4x^2 - 12x + 8 = 0$

$4x - 4$ $x_1 = 1, x_2 = 2$
$x - 2 = 4x^2 - 12x + 8$

The left side of the equation is factorised and $4x^2$ is split into 4x/x and 8 into 2/4. Put the factors under each other and provide the factors 2/4 so the result after crosswise multiplication is the same as the polynomial.

Example 19 $5x^2 + 17x + 14 = 0$

$5x + 7$ $x_1 = -7/5, x_2 = -2$
$x + 2 = 5x^2 + 17x + 14$

The left side of the equation is factorised and $5x^2$ is split into 5x/x and 14 into 2/7. Put the factors under each other and provide the factors 2/4 so the result after crosswise multiplication is the same as the polynomial. Put the factors under each other and provide the factors 2/4 so the result after crosswise multiplication is the same as the polynomial.

Example 20 $2x^2 - 4x - 30 = 0$

$2x + 6$ $x_1 = -3, x_2 = 5$
$x - 5 = 2x^2 - 4x - 30$

The left side of the equation is factorised and $2x^2$ is split into 2x/x and 30 into 5/6. Put the factors under each other and provide the factors 5/6 so the result after crosswise multiplication is the same as the polynomial.

Example 21 $2x^2 + 7x + 5 = 0$

$2x + 5$ $x_1 = -5/2, x_2 = -1$
$x + 1 = 2x^2 + 7x + 5$

The left side of the equation is factorised and $2x^2$ is split into 2x/x and 5 into 5/1. Put the factors under each other and provide the factors 5/1 so the result after crosswise multiplication is the same as the polynomial.

In fact whenever the outer coefficients add up to the middle coefficient (x+1) is a factor.

Example 22 $3x^2-10x-8=0$

$3x+2$ $x_1 = -2/3, \ x_2 = 4$

$x - 4 = 3x^2 - 10x - 8$

The left side of the equation is factorised and $3x^2$ is split into $3x/x$ and 8 into 2/4. Put the factors under each other and provide the factors 2/4 so the result after crosswise multiplication is the same as the polynomial.

Example 23 $8x^2 + 2x-15=0$

$4x - 5$ $x_1 = 5/4, \ x_2 = -3/2$

$2x + 3 = 8x^2 + 2x - 15$

The left side of the equation is factorised and $8x^2$ is split into $4x/2x$ and 15 into 3/5. Put the factors under each other and provide the factors 3/5 so the result after crosswise multiplication is the same as the polynomial.

Example 24 $3x^2 + 5x-8=0$

$3x + 8$ $x_1 = -8/3, \ x_2 = 1$

$x - 1 = 3x^2 + 5x - 8$

The left side of the equation is factorised and $3x^2$ is split into $3x/x$ and 8 into 8/1 since the middle coefficient 5 is an odd number. Put the factors under each other and provide the factors 2/4 so the result after crosswise multiplication is the same as the polynomial.

Example 25 $6x^2 - 7x - 3=0$

$3x + 1$ $x_1 = -1/3, \ x_2 = 3/2$

$2x - 3 = 6x^2 - 7x - 3$

The left side of the equation is factorised and $6x^2$ is split into $3x/2x$ and 3 into 3/1. Put the factors under each other and provide the factors 1/3 so the result after crosswise multiplication is the same as the polynomial.

Example 26 $2x^2 + x - 6 = 0$

$2x - 3$ $\qquad\qquad\qquad\qquad$ $x_1 = 3/2,\ x_2 = -2$

$x + 2 = 2x^2 + x - 6$

The left side of the equation is factorised and $2x^2$ is split into 2x/x and 6 into 3/2. Put the factors under each other and provide the factors 1/3 so the result after crosswise multiplication is the same as the polynomial.

Example 27 $2x^2 + 11x + 12 = 0$

$2x + 3$ $\qquad\qquad\qquad\qquad$ $x_1 = -3/2,\ x_2 = -4$

$x + 4 = 2x^2 + 11x + 12$

The left side of the equation is factorised and $2x^2$ is split into 2x/x and 12 into ¾. Put the factors under each other and provide the factors 1/3 so the result after crosswise multiplication is the same as the polynomial.

Example 28 $2x^2 + 9x + 4 = 0$

$2x + 1$ $\qquad\qquad\qquad\qquad$ $x_1 = -1/2,\ x_2 = -4$

$x + 4 = 2x^2 + 9x + 4$

The left side of the equation is factorised and $2x^2$ is split into 2x/x and 4 into 4/1. Put the factors under each other and provide the factors 1/4 so the result after crosswise multiplication is the same as the polynomial.

Example 29 $6x^2 + 17x + 12 = 0$

$3x + 4$ $\qquad\qquad\qquad\qquad$ $x_1 = -4/3,\ x_2 = -3/2$

$2x + 3 = 6x^2 + 17x + 12$

The left side of the equation is factorised and $6x^2$ is split into 3x/2x and 12 into 4/3. Put the factors under each other and provide the factors 3/4 so the result after crosswise multiplication is the same as the polynomial.

Example 30 $4x^2 + 15x + 9 = 0$

4x + 3 $x_1 = -3/4$, $x_2 = -3$

  x + 3 = $4x^2 + 15x + 9$

The left side of the equation is factorised and $4x^2$ is split into 4x/x and 9 into 3/3. Put the factors under each other and provide the factors 3/3 so the result after crosswise multiplication is the same as the polynomial.

Example 31 $3x^2 + 13x + 12 = 0$

3x + 4 $x_1 = -4/3$, $x_2 = -3$

  x + 3 = $3x^2 + 13x + 12$

The left side of the equation is factorised and $3x^2$ is split into 3x/x and 12 into 4/3. Put the factors under each other and provide the factors 3/4 so the result after crosswise multiplication is the same as the polynomial.

Example 32 $12x^2 + 13x + 3 = 0$

4x + 3 $x_1 = -3/4$, $x_2 = -1/3$

3x + 1 = $12x^2 + 13x + 3$

The left side of the equation is factorised and $12x^2$ is split into 4x/3x and 3 into 3/1. Put the factors under each other and provide the factors 3/3 so the result after crosswise multiplication is the same as the polynomial.

Example 33 $5x^2 + 16x + 12 = 0$

5x + 6 $x_1 = -6/5$, $x_2 = -2$

  x + 2 = $5x^2 + 16x + 12$

The left side of the equation is factorised and $5x^2$ is split into 5x/x and 12 into 2/6. Put the factors under each other and provide the factors 2/6 so the result after crosswise multiplication is the same as the polynomial.

Example 34 $3x^2 + 11x + 6 = 0$

$3x + 2$ $x_1 = -2/3, x_2 = -3$

$x + 3 = 3x^2 + 11x + 6$

The left side of the equation is factorised and $3x^2$ is split into 3x/x and 6 into 3/2. Put the factors under each other and provide the factors 2/3 so the result after crosswise multiplication is the same as the polynomial.

Example 35 $6x^2 + 13x + 6 = 0$

$3x + 2$ $x_1 = -2/3, x_2 = -3/2$

$2x + 3 = 6x^2 + 13x + 6$

The left side of the equation is factorised and $6x^2$ is split into 3x/2x and 6 into 3/2. Put the factors under each other and provide the factors 2/3 so the result after crosswise multiplication is the same as the polynomial.

Example 36 $7x^2 - 10x - 8 = 0$

$7x + 4$ $x_1 = -4/7, x_2 = 2$

$x - 2 = 7x^2 - 10x - 8$

The left side of the equation is factorised and $7x^2$ is split into7x/x and 8 into 2/4. Put the factors under each other and provide the factors 2/4 so the result after crosswise multiplication is the same as the polynomial.

Example 37 $5x^2 + 18x - 8 = 0$

$5x - 2$ $x_1 = 2/5, x_2 = -4$

$x + 4 = 5x^2 + 18x - 8$

The left side of the equation is factorised and $5x^2$ is split into 5x/x and 8 into 2/4. Put the factors under each other and provide the factors 2/4 so the result after crosswise multiplication is the same as the polynomial.

Example 38 $3x^2 + 10x + 8 = 0$

$3x + 4$ $x_1 = -4/3, x_2 = -2$

$x + 2 = 3x^2 + 10x + 8$

The left side of the equation is factorised and $3x^2$ is split into 3x/x and 8 into 2/4. Put the factors under each other and provide the factors 2/4 so the result after crosswise multiplication is the same as the polynomial.

Example 39 $7x^2 + 8x - 12 = 0$

$7x - 6$ $x_1 = 6/7, \; x_2 = -2$

$x + 2 = 7x^2 + 8x - 12$

The left side of the equation is factorised and $7x^2$ is split into 7x/x and 12 into 2/6. Put the factors under each other and provide the factors 2/6 so the result after crosswise multiplication is the same as the polynomial.

Example 40 $12x^2 + 33x + 18 = 0$

$4x + 3$ $x_1 = -3/4, \; x_2 = -2$

$3x + 6 = 12x^2 + 33x + 18$

The left side of the equation is factorised and $12x^2$ is split into 4x/3x and 18 into 3/6. Put the factors under each other and provide the factors 3/6 so the result after crosswise multiplication is the same as the polynomial.

Multiplication of numbers

In the same way as algebraic expressions are multiplied, numbers can also be multiplied and that is done in one line from right to left or from left to right.

Example 1   42 x 31

42
31
<u>13₁0 2</u>

We perform three operations:

1. Multiply the right-hand figures of the two numbers and put their product down as the right-hand figure of the answer: 2 x 1 = 2.

2. Cross-multiply and add: (4 x 1) + (2 x 3) = 10, put down 0 and carry 1, as shown.

3) Multiply the left-hand figures: 4 x 3 = 12, add the carried 1 and put down 13.

So 42 x 31 = 1302, the product being achieved in one line by multiplying vertically on the right, crosswise, and then vertically on the left.

It is important to see the vertically and crosswise pattern in this calculation, so that we can easily extend it to larger products. If we put 4 dots to represent the digits in the sum, then the 3 steps can be illustrated:

3rd step	2nd step	1st step
4 x 3 = 12	(4 x 1) + (2 x 3) = 10	2 x 1= 2

Explanation

We multiply units by units to get the units digit of the answer, tens by units and units by tens to get the tens digit of the answer, and tens by tens to get the hundreds digit of the answer. The method can be extended to apply to numbers of any size.

Another way of multiplying

In Vedic Mathematics mental calculation played a central role in mathematical calculations, as they probably had no access to pen or paper. They made use of many tricks, e.g. bases, 1, 10, 100, etc. A few examples from Vedic Mathematics illustrate the way these methods were used. There is only one conventional method available for multiplication and it is especially cumbersome when the digits of the numbers being multiplied are large, for example 88 x 98. The formula "All from 9 and the Last from ten", enables us to make use of the fact that the numbers are near to a base, and give the answer easily and immediately as the following examples show.

Example 2   88 x 98

88 –12   Both of these numbers are close to a base of 100. We first write
98 – 2   the numbers under each other, and write their deficiencies from
86 / 24   100 on the right. These deficiencies are prefixed by a minus sign
as the numbers being multiplied are below the base, and they are obtained simply by observation or by applying the formula "All from 9 and the Last from 10".

The left part of the answer is obtained by cross-subtraction, 88– 02 = 86 or 98 –12 = 86. The right part is obtained by multiplying the deviations i.e. 12 x 2 = 24. This method is so easy that it is simple to do the whole calculation mentally: we just subtract the deficiency of one number from the other number, to get the first part of the answer, and then multiply the deficiencies to get the last part.

Explanation   88 x 98 = 88 x 100 – 88 x 2
= 8800 – (100 x 2 – 12 x 2)
= 8800 – 200 + 12 x 2
= 8600 + 24 = 8624.

Example 3   **9 x 8**

$$9 - 1$$
$$\underline{8 - 2}$$
$$7 / 2$$

We use a base of 10. As can be seen from this example, multiplication tables above 5 x 5 are not really needed in Vedic Mathematics. All we multiply here are the deficiencies 1 and 2.

Example 4   **67857 x 99998**

$$67859 - 32141$$
$$\underline{99998 - \quad\quad 2}$$
$$67857 \ / 64282$$

Here the base used is 100 000. The deficiencies are quickly obtained by All from 9 and the Last from 10.

Example 5   **97 x 98**

$$97 - 3$$
$$\underline{98 - 2}$$
$$95/ 06$$

For a base 100 there must be two figures on the RHS. We therefore insert a zero before the 6.

Example 6   **112 x 97**

$$112 + 12$$
$$\underline{97 - 03}$$
$$109 / \overline{36}$$
$$108 / 64$$

One deficiency is positive and one negative. Therefore for the left hand side, LHS 112– 3 = 109 or 97 + 12= 109. For the right hand side RHS the product of the deficiencies is negative i.e. 36. We therefore reduce the LHS by 1 and apply the formula to the RHS. (3 from 9 and 6 from 10) We are actually subtracting 36 from one of the hundreds in the third column.

Example 7   98 x 97 x 96
$$98 - 02$$
$$97 - 03$$
$$\underline{96 - 04}$$
$$91/\,26\,/\,\overline{24}$$
$$\underline{91\,/25\,/\,76}$$

Again we put the deficiencies next to the numbers. For the LHS of the answer subtract: $98 - 3 - 4 = 91$, $97 - 2 - 4 = 91$ or $96 - 2 - 3 = 91$.
For the middle part multiply the deficiencies in pairs and add,
i.e. $(2 \times 3) + (2 \times 4) + (3 \times 4) = 26$.
For the RHS multiply the deficiencies, but as there are three, their product is negative.

Example 8   9111 x 9900

$$9111 - 0889$$
$$\underline{9900 - 0100}$$
$$9011\,/_8 8900$$
$$\underline{9019\,/\,8900}$$

When the numbers being multiplied are both below the base there will be as many figures on the RHS as there are on the LHS; we can add zeros to the deficiencies if we wish to maintain this symmetry.

Example 9   99998 x 96

$$99998\ - 00002$$
$$\underline{96\qquad - 04}$$
$$95998\,/\ \ 08$$

Here we should put the smaller number, and its deficiency, as far to the left as possible as shown.

Example 10 $75^2 = 56,25$

When squaring of numbers ending or beginning in five an old formula is used, By one more than one before and it is also a sutra. By one more than the one before means here, multiply 7 by 8 i.e. multiply 7 by the number following 7 and RHS is simply 25, so $75^2 = 56,25$.

Example 11 $72 \times 78 = 56,16$

The same formula By one more than the one before applies to the product of two numbers with the same first digit and where the last digits are total 10.
Here both numbers begin with 7 and $2 + 8 = 10$.
So as example 10 the first part of the answer is 56.
The last part is $2 \times 8 = 16$ so $72 \times 78 = 56,16$.

Chapter 3

CUBIC EQUATIONS

A cubic equation has all information about its origin in the coefficient of the x^2 term, in the same manner as the middle coefficient in a quadratic equation.

When the first coefficient is 1, the origin of the equation is visible in the coefficient of the x^2-term, and therefore such equations are easy to solve.

But if the first coefficient is larger than 1, the origin of the equation is not visible in the coefficient of the x^2-term. Therefore such an equation has to be solved first by testing factors or by using the factor rule. Since the origin of the equation is not visible in such an equation, the origin will be illustrated in the two first examples.

The coefficients are split into factors

The table below shows how the coefficients are split into factors

1. 1,1.1	2. 1,1,2	3. 1,1,3
4. 1,1,4/ 1,2,2	5. 1,1,5	6. 1,2,3 /1,1,6
7. 1,1,7	8. 1,2,4 /1,1,8	9. 1,3,3 /1,1,9
10. 1,2,5/ 1,1,10	11. 1,1,11	12. 1,3,4 / 1,2,6 / 1,1,12
13. 1,1,13	14. 1,2,7/ 1,1,14	15. 1,3,5 / 1,1,15
16. 1,2,8/1,4,4 ,/ 2,2,4 /1,1,16	17. 1,1,17	18. 1,2,9/13,6/ 1,1,18
19. 1,1,19	20. 2,2,5/ 1,4,5 /1,1,20	21. 1,3,7 / 1,1,21
22. 1,2,11/1,1,22	23. 1,1,23	24. 2,3,4 / 1,3,8 /1,1,24

Equations when the first coefficient is 1

If the first coefficient is one the coefficient of the x^2-term shows the sum of the factors, and the last term shows the product of the factors.

Example 1 $x^3 + 7x^{2*} + 14x + 8 = 0$ $(1+2+4 = +7x^2)^{1)}$

The left side of the equation is factorised and x^3 is split up into $x/x/x$, and 8 is split into 1/2/4. The factors are $(x + 1)(x + 2)(x + 4)$ and the sum of them is 7, like the coefficient of the x^2-term. The product of the factors is like the last term 8. Cross-multiplication is at the same time a way to check the answer.

$x^3 + 7x^2 + 14x + 8$
$x + 1$
$x + 2 = x^2 + 3x + 2$
         $x + 4 = x^3 + 7x^2 + 14x + 8$

Each factor is put to zero, e.g. $x + 1 = 0$, $x = -1$
$x_1 = -1$, $x_2 = -2$, $x_3 = -4$

1st step: $x \cdot x = x^2$
2nd step: $x \cdot 2 + x \cdot 1 = 3x$
3rd step: $1 \cdot 2 = 2$
4th step: $x^2 \cdot x = x^3$
5th step: $x^2 \cdot 4 + 3x \cdot x = 7x^2$
6th step: $3x \cdot 4 + x \cdot 2 = 14x$
7th step: $2 \cdot 4 = 8$

Prime factorising is especially suitable for large numbers, see page 82.

*the origin of the equation

[1] it shows the whole solution in a nutshell.

Example 2 $x^3 + 9x^2 + 24x + 20 = 0$ $\qquad\qquad$ $(2+2+5= 9x^2)$

The left side of the equation is factorised and x^3 is split up into x/x/x and 20 is factorised and we get; $20 = 2 \cdot 10 = 2 \cdot 2 \cdot 5$. The sum of the factors is like the coefficient of the x^2-term and the product is 20. The factors are $(x + 2)(x + 2)(x + 5)$ as shown below.

$x^3 + 9x^2 + 24x + 20$

$x + 2$

$x + 2 = x^2 + 4x + 4$

$\qquad\qquad x + 5 = x^3 + 9x^2 + 24x + 20$

$x_1 = x_2 = -2, x_3 = -5$

When the numerators $2 \cdot 2 \cdot 5$ are multiplied we get 20 like the last term, and when the denominators $1 \cdot 1 \cdot 1$ are multiplied we get 1 like the first coefficient x^3.

Example of the graph to the third degree function: $y = x^3 + 9x^2 + 24x + 20$

a)

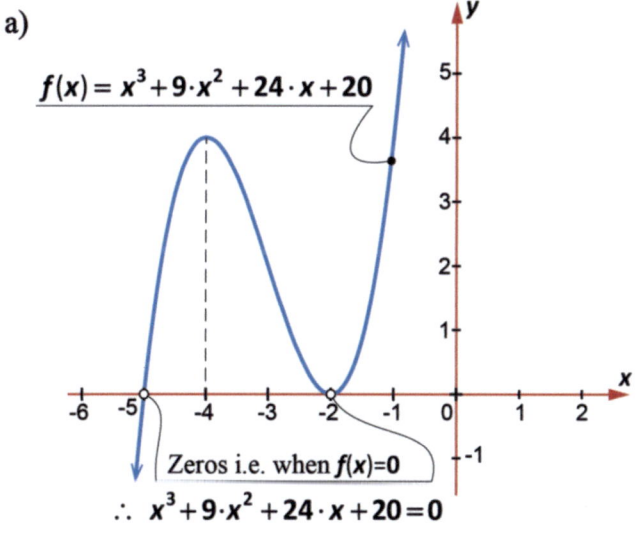

$f(x) = x^3 + 9 \cdot x^2 + 24 \cdot x + 20$

Zeros i.e. when $f(x)=0$

$\therefore x^3 + 9 \cdot x^2 + 24 \cdot x + 20 = 0$

b)

x	$f(x)$	(x, y)
-6	-16	-6, 14
-5	0	-6, 0
-4	4	-4, 4
-3	2	-3, 2
-2	0	-2, 0
-1	4	-1, 4
0	20	0, 20
1	54	1, 54

Comparison with a conventional method

Polynomial division

$x^3 + 9x^2 + 24x + 20 = 0$

By polynomial division the equation is broken down into a quadratic equation, provided you know the root, or if nothing else, by a guess or by a test. We found that $x = -2$.

$x^2 + 7x + 10$

$x^3 + 9x^2 + 24x + 20/x + 2$

$\underline{x^3 + 2\,x^2}$ $x^2\,(x + 2)$

      $7\,x^2 + 24x$ $7x\,\,(x + 2)$

      $\underline{7x^2 + 14x}$ $10\,\,(x + 2)$

            $10x + 20$

            $\underline{10\,x + 20}$

                0

Quadratic equation $x^2 + 7x + 10 = 0$ is solved as follows:

$x = -3{,}5 \pm \sqrt{12{,}25 + 1(-10)} = -3{,}5 \pm \sqrt{2{,}25} = -3{,}5 \pm 1{,}5$

$x_1 = x_2 = -2, \;\; x_3 = -5$

Example 3 $x^3 + 11x^2 + 38x + 40 = 0$ $(2+4+5 = 11x^2)$

The left side of the equation is factorised and x^3 is split to x/x/x and 40 is factorised and we get: $40 = 2 \cdot 20 = 2 \cdot 4 \cdot 5$. The sum of the factors is 11 like the coefficient of the x^2-term and the product is 40. The factors are $(x + 2)(x + 4)(x +5)$ as shown below.

$x^3 + 11x^2 + 38x + 40$

$x + 2$

$x + 4 = x^2 + 6x + 8$

         $x + 5 = x^3 + 11x^2 + 38x + 40$

$x_1 = -2, \; x_2 = -4, \; x_3 = -5$

Example 4 $x^3 - 23x^2 + 142x - 120 = 0$ $\qquad\qquad$ $(-1-10-12 = -23x^2)$

The left side of the equation is factorised and x^3 is split into x/x/x and 120 into 1/10/12. The sum of the factors is -23 like the coefficient of the x^2-term and the product is -120. The factors are $(x - 1)(x - 10)(x - 12)$ as shown below.

$x^3 - 23x^2 + 142x - 120$
$x - 1$
$x - 10 = x^2 - 11x + 10$
$\qquad\quad x - 12 = x^3 - 23x^2 + 142x - 120$
$x_1 = 1, x_2 = 10, \; x_3 = 12$

Example 5 $x^3 + 15x^2 + 71x + 105 = 0$ $\qquad\qquad$ $(3+5+7=15\ x^2)$

The left side of the equation is factorised and x^3 is split into x/x/x and 105 is factorised and we get: $105 = 3 \cdot 35 = 3 \cdot 5 \cdot 7$. The sum of the factors is 15 as the coefficient of the x^2-term and the product is 105. The factors are $(x + 3)(x + 5)(x + 7)$ as follows.

$x^3 + 15x^2 + 71x + 105$
$x + 3$
$x + 5 = x^2 + 8x + 15$
$\qquad\quad x + 7 = x^3 + 15x^2 + 71x + 105$

$x_1 = -3, x_2 = -5, x_3 = -7$

Example 6 $x^3 + 6x^2 + 11x + 6 = 0$ $\qquad\qquad$ $(1+2+3= 6\ x^2)$

The left side of the equation is factorised and x^3 is split into x/x/x and 6 is split to 1/2/3. The coefficient of the x^2-term is 6 and the product is 6. The factors $(x + 1)(x + 2)(x + 3)$ are as shown below.

$x^3 + 6x^2 + 11x + 6$
$x + 1$
$x + 2 = x^2 + 3x + 2$
$\qquad\quad x + 3 - x^3 + 6x^2 + 11x + 6$

$x_1 = -1, x_2 = -2, x_3 = -3$

Example 7 $x^3 - 2x^2 - 5x + 6 = 0$ $(-1+2-3 = -2x^2)$

The left side of the equation is factorised and x^3 is split into x/x/x and 6 is split to 1/2/3.
The coefficient of the x^2-term is -2 and the product is 6. The factors $(x - 1)(x + 2)(x - 3)$
are as shown below.

$x^3 - 2x^2 - 5x + 6$

x $-$ 1

x + 2 $= x^2 + x - 2$

$\qquad$ x $-$ 3 $= x^3 - 2x^2 - 5x + 6$

$x_1 = 1, x_2 = -2, x_3 = 3$

Example 8 $x^3 + 3x^2 - 4x - 12 = 0$ $(2+3-2 = +3x^2)$

The left side of the equation is factorised and x^3 is split to x/x/x and 12 is factorised and
we get: $12 = 2 \cdot 6 = 2 \cdot 3 \cdot 2$. The coefficient of the x^2-term is 3 and the product is -12.
The factors are $(x + 2)(x + 3)(x - 2)$ as shown below.

$x^3 + 3x^2 - 4x - 12$

x + 2

x + 3 $= x^2 + 5x + 6$

$\qquad$ x $-$ 2 $= x^3 + 3x^2 - 4x - 12$

$x_1 = -3, x_2 = -2, x_3 = 2$

Example 9 $x^3 + 9x^2 + 26x + 24 = 0$ $(3+2+4 = 9x^2)$

The left side of the equation is factorised and x^3 is split to x/x/x and 24 is factorised and
we get: $24 = 3 \cdot 8 = 3 \cdot 2 \cdot 4$. The coefficient of the x^2-term is 9 and the product is 24.
The factors are $(x + 3)(x + 2)(x + 4)$ as shown below.

$x^3 + 9x^2 + 26x + 24$

x + 3

x + 2 $= x^2 + 5x + 6$

$\qquad$ x + 4 $= x^3 + 9x^2 + 26x + 24$

$x_1 = -3, x_2 = -2, x_3 = -4$

Example 10 $x^3 + 10x^2 + 27x + 18 = 0$ $(1+3+6 = +10x^2)$

The left side of the equation is factorised and x^3 is split into x/x/x and 18 into 1/3/6. The coefficient of the x^2-term is 10 and the product is 18. The factors $(x + 1)(x+ 3)(x + 6)$ are as shown below.

$x^3 + 10x^2 + 27x + 18$

$x + 1$

$x + 3 = x^2 + 4x + 3$

$\qquad\qquad x + 6 = x^3 + 10x^2 + 27x + 18$

$x_1 = - 1, x_2 = - 3, x_3 = - 6$

Example 11 $x^3 + 9x^2 + 23x + 15 = 0$ $(1+3+5 = + 9x^2)$

The left side of the equation is factorised and x^3 is split into x/x/x and 15 into 1/3/5. The coefficient of the x^2-term is 9 and the product is 15. The factors $(x + 1)(x + 3)(x + 5)$ are as shown below.

$x^3 + 9x^2 + 23x + 15$

$x + 1$

$x + 3 = x^2 + 4x + 3$

$\qquad\qquad x + 5 = x^3 + 9x^2 + 23x + 15$

$x_1 = - 1, x_2 = - 3, x_3 = - 5$

Example 12 $x^3 + 10x^2 + 29x + 20 = 0$ $(1+4+5= +10x^2)$

The left side of the equation is factorised and x^3 is split into x/x/x and 20 is split 1/4/5. The coefficient of the x^2-term is 10 and the product is 20. As we have shown below the factors are $(x + 1)(x + 4)(x +5)$.

$x^3 + 10x^2 + 29x + 20$

$x + 1$

$x + 4 = x^2 + 5x + 4$

$\qquad\qquad x + 5 = x^3 + 10x^2 + 29x + 20$

$x_1 = - 1, x_2 = - 4, x_3 = - 5$

Example 13 $x^3 + 2x^2 - 11x - 12 = 0$ $(1+4-3 = +2\ x^2)$

The left side of the equation is factorised and x^3 is split into x/x/x and 12 is split 1/4/3.
The coefficient of the x^2-term is 2 and the product is –12. As we have shown below the
factors are $(x + 1)(x + 4)(x - 3)$.
$x^3 + 2x^2 - 11x - 12$
$x + 1$
$x + 4 = x^2 + 5x + 4$
$\qquad\qquad x - 3 = x^3 + 2x^2 - 11x - 12$
$x_1 = -1,\ x_2 = -4,\ x_3 = 3$

Example 14 $x^3 + 4x^2 + x - 6 = 0$ $(2+3-1= +4x^2)$

The left side of the equation is factorised and x^3 is split into x/x/x and 6 is split to 2/3/1.
The coefficient of the x^2-term is 4 and the product is– 6. The factors $(x + 2)(x + 3)(x -1)$
are as shown below.
$x^3 + 4x^2 + x - 6$
$x + 2$
$x + 3 = x^2 + 5x + 6$
$\qquad\qquad x - 1 = x^3 + 4x^2 + x - 6$

$x_1 = -2,\ x_2 = -3,\ x_3 = 1$

The steps are as follows:
1:st step: $x \cdot x = x^2$
2:nd step: $x \cdot 3 + x \cdot 2 = 5x$
3:rd step: $2 \cdot 3 = 6$
4:th step: $x^2 \cdot x = x^3$
5:th step: $x^2 (-1) + 5x \cdot x = 4x^2$
6:th step: $6 \cdot x + 5x(-1) = x$
7:th step: $6(-1) = -6$

Example 15 $x^3 - 3x^2 - 6x + 8 = 0$ $\qquad$ $(-4 -1+2 = -3x^2)$

The left side of the equation is factorised and x^3 is split into x/x/x and 8 is split to 4/1/2. The coefficient of the x^2-term is –3 and the product is 8. The factors $(x - 4)(x -1)(x + 2)$ are as shown below.

$x^3 - 3x^2 - 6x + 8$

$x - 4$

$x - 1 = x^2 - 5x + 4$

$\qquad\quad x + 2 = x^3 - 3x^2 - 6x + 8$

$x_1 = 4,\ x_2 = 1,\ x_3 = -2$

Example 16 $x^3 + 3x^2 - 6x - 8 = 0$ $\qquad$ $(+4+1-2 = +3x^2)$

The left side of the equation is factorised and x^3 is split into x/x/x and 8 is split to 4/1/2. The coefficient of the x^2-term is 3 and the product is–8. The factors $(x + 4)(x + 1)(x - 2)$ are as shown below.

$x^3 + 3x^2 - 6x - 8$

$x + 4$

$x + 1 = x^2 + 5x + 4$

$\qquad\quad x - 2 = x^3 + 3x^2 - 6x - 8$

$x_1 = -4,\ x_2 = -1,\ x_3 = 2$

Example 17 $x^3 - 7x^2 + 4x + 12 = 0$ $\qquad$ $(-2- 6+1= -7x^2)$

The left side of the equation is factorised and x^3 is split into x/x/x and 12 is split 2/6/1. The coefficient of the x^2-term is –7 and the product is 12. As we have shown below the factors are $(x - 2)(x - 6)(x + 1)$.

$x^3 - 7x^2 + 4x + 12$

$x - 2$

$x - 6 = x^2 - 8x + 12$

$\qquad\quad x + 1 = x^3 - 7x^2 + 4x + 12$

$x_1 = 2,\ x_2 = 6,\ x_3 = -1$

Example 18 $x^3 - 6x^2 + 11x - 6 = 0$ $\hspace{2cm}$ $(-1-2-3 = -6x^2)$

The left side of the equation is factorised and x^3 is split into x/x/x and 6 to 1/2/3. The coefficient of the x^2-term is–6 and the product is–6. The factors $(x - 1)(x - 2)(x - 3)$ are as shown below.

$x^3 - 6x^2 + 11x - 6$

$x - 1$

$x - 2 = x^2 - 3x + 2$

$\hspace{3cm} x - 3 = x^3 - 6x^2 + 11x - 6$

$x_1 = 1,\ x_2 = 2,\ x_3 = 3$

Example 19 $x^3 - 2x^2 - x + 2 = 0$ $\hspace{2cm}$ $(-2-1+1 = -2x^2)$

The left side of the equation is factorised and x^3 is split into x/x/x and 2 is split to 2/1/1. The coefficient of the x^2-term is–2 and the product is 2. The factors $(x - 2)(x - 1)(x + 1)$ are as shown below.

$x^3 - 2x^2 - x + 2$

$x - 2$

$x - 1 = x^2 - 3x + 2$

$\hspace{3cm} x + 1 = x^3 - 2x^2 - x + 2$

$x_1 = 2,\ x_2 = 1,\ x_3 = -1$

Example 20 $x^3 + 2x^2 - 11x - 12 = 0$ $\hspace{2cm}$ $(1 - 3 + 4 = +2x^2)$

The left side of the equation is factorised and x^3 is split to x/x/x and 12 is split to 1/3/4. The coefficient of the x^2-term is 2 and the product is– 12. The factors $(x +1)(x-3)(x+ 4)$ are as shown below.

$x^3 + 2x^2 - 11x - 12$

$x + 1.$

$x - 3 = x^2 - 2x - 3$

$\hspace{3cm} x + 4 = x^3 + 2x^2 - 11x - 12$

$x_1 = -1,\ x_2 = 3,\ x_3 = -4$

Example 21 $x^3 + 5x^2 + 2x - 8 = 0$ $(-1+2+4 = +5x^2)$

The left side of the equation is factorised and x^3 is split into x/x/x and 8 is split to 1/2/4. The coefficient of the x^2-term is 5 and the product is–8. The factors $(x - 1)(x + 2)(x + 4)$ are as shown below.

$x^3 + 5x^2 + 2x - 8$

$x - 1$

$x + 2 = x^2 + x - 2$

$\qquad x + 4 = x^3 + 5x^2 + 2x - 8$

$x_1 = 1, \; x_2 = -2, \; x_3 = -4$

Example 22 $x^3 - 4x^2 - 9x + 36 = 0$ $(-3-4+3 = -4x^2)$

The left side of the equation is factorised and x^3 is split to x/x/x and 36 is split to 3/4/3. The coefficient of the x^2-term is–4 and the product is 36. The factors $(x - 3)(x - 4)(x + 3)$ are as shown below.

$x^3 - 4x^2 - 9x + 36$

$x - 3$

$x - 4 = x^2 - 7x + 12$

$\qquad x + 3 = x^3 - 4x^2 - 9x + 36$

$x_1 = 3, \; x_2 = 4, \; x_3 = -3$

Example 23 $x^3 + 8x^2 + 19x + 12 = 0$ $(4+3+1 = +8x^2)$

The left side of the equation is factorised and x^3 is split into x/x/x and 12 into 4/3/1. The coefficient of the x^2-term is 8 and the product is 12. The factors $(x + 4)(x + 3)(x + 1)$ are as shown below.

$(x + 4)(x + 3)(x + 1)$

$x + 4$

$x + 3 = x^2 + 7x + 12$

$\qquad x + 1 = x^3 + 8x^2 + 19x + 12$

$x_1 = -4, \; x_2 = -3, \; x_3 = -1$

Example 24 $x^3 + 5x^2 - 2x - 8 = 0$ $(+4+2-1= +5x^2)$

The left side of the equation is factorised and x^3 is split into x/x/x and 8 is split to 4/2/1. The coefficient of the x^2-term is 5 and the product is –8. The factors (x +4)(x +2) (x – 1) are as shown below.

$x^3 + 5x^2 + 2x - 8$

x + 4

$x + 2 = x^2 + 6x + 8$

$\qquad\qquad x - 1 = x^3 + 5x^2 + 2x - 8$

$x_1 = -4, \; x_2 = -2, \; x_3 = 1$

Example 25 $x^3 - 5x^2 - 2x + 24 = 0$ $(-3- 4+2 = -5x^2)$

The left side of the equation is factorised and x^3 is split into x/x/x and 24 into 3/4/2. The coefficient of the x^2-term is–5 and the product is 24. The factors (x – 3)(x – 4)(x + 2) are as shown below.

$x^3 - 5x^2 - 2x + 24$

x - 3

$x - 4 = x^2 - 7x + 12$

$\qquad\qquad x + 2 = x^3 - 5x^2 - 2x + 24$

$x_1 = 3, \; x_2 = 4, \; x_3 = -2$

Example 26 $x^3 + 3x^2 - x - 3 = 0$ $(+3+1-1= +3x^2)$

The left side of the equation is factorised and x^3 is split into x/x/x and 3 is split to 3/1/1. The coefficient of the x^2-term is 3 and the product is –3. The factors (x + 3)(x+ 1)(x – 1) are as shown below.

$x^3 + 3x^2 - x - 3$

x + 3

$x + 1 = x^2 + 4x + 3$

$\qquad\qquad x - 1 = x^3 + 3x^2 - x - 3$

$x_1 = -3, \; x_2 = -1, \; x_3 = 1$

Example 27 $x^3 + 4x^2 - 3x - 18 = 0$ $(-2+3+3=+4x^2)$

The left side of the equation is factorised and x^3 is split to $x/x/x$ and 18 is split to 2/3/3. The coefficient of the x^2-term is 4 and the product is–18. The factors $(x-2)(x+3)(x+3)$ are as shown below.

$x^3 + 4x^2 - 3x - 18$

$x - 2$

$x + 3 = x^2 + x - 6$

$\quad\quad\quad x + 3 = x^3 + 4x^2 - 3x - 18$

$x_1 = 2$, $x_2 = -3$, $x_3 = -3$

Example 28 $x^3 + 6x^2 + 3x - 10 = 0$ $(5+2-1=+6x^2)$

The left side of the equation is factorised and x^3 is split into $x/x/x$ and 10 is split 5/2/1. The coefficient of the x^2-term is 6 and the product is–10. The factors $(x+5)(x+2)(x-1)$ are as shown below.

$x^3 + 6x^2 + 3x - 10$

$x + 5$

$x + 2 = x^2 + 7x + 10$

$\quad\quad\quad x - 1 = x^3 + 6x^2 + 3x - 10$

$x_1 = -5$, $x_2 = -2$, $x_3 = 1$

Equations when the first coefficient is larger than 1

When the first coefficient is larger than 1, the origin of the equation is not visible in the coefficient of the x^2-term. One way to solve such an equation is to test the factors first. If two factors are found, the third one is also given when it is provided with a plus or a minus sign. If no factors are found: "Multiply the largest coefficient in the first term by the largest factor in the last term."

Example 29 $3x^3 - 16x^2 + 23x - 6 = 0$

The left side of the equation is factorised and $3x^3$ is split into $3x/x/x$ and 6 is into 1/2/3. A test showed that both 2 and 3 are a root and we get the factors $(x - 2)$ and $(x - 3)$.

If 2 is a root we get: $3 \cdot 2^3 = 24 - 16 \cdot 2^2 = -64 + 23 \cdot 2 = 46 - 6$, i.e. $24 - 64 + 46 - 6 = 0$.

If 3 is a root we get; $3 \cdot 3^3 = 81 - 16 \cdot 3^2 = -144 + 23 \cdot 3 = 69 - 6$, i.e. $81 - 144 + 69 - 6 = 0$.

The third factor only needs to be provided with a minus sign $(3x - 1)$ as shown below.

$3x^3 - 16x^2 + 23x - 6$

$x - 2$

$x - 3 = x^2 - 5x + 6$

$\qquad\qquad 3x - 1 = 3x^3 - 16x^2 + 23x - 6$

$x_1 = 2, x_2 = 3$ $x_3 = 1/3$

To show that the coefficient of the x^2-term 16 is the origin of the equation, the first coefficient of the polynomial is divided by 3 as shown below.

$3x^3 - 16x^2 + 23x - 6$

$\dfrac{3}{3}x^3 - \dfrac{16}{3}x^2 + \dfrac{23}{3}x - \dfrac{6}{3}$

$x^2 - \dfrac{16}{3}x^2 + \dfrac{23}{3}x - \dfrac{6}{3}$

$\qquad\quad 5, 33 \qquad\qquad 2$

When the answer is multiplied $2 \cdot 3 \cdot 1/3$ we get 2, as the last term 6/3, and when we add the answer $1/3 + 2 + 3$ we get 5.33 corresponding to 16/3.

Example 30 $15x^3 - 31x^2 + 0x + 4 = 0$

The left side of the equation is factorised and $15x^3$ is split into $5x/3x/x$ and 4 is split into 2/2/1 since $(x-2)$ is a factor. We get the second factor when the largest coefficient $5x$ is multiplied by the largest factor 2 in the last term. A swift glance at $-31x^2$ indicates that the factor is likely $(5x-2)$. After multiplication of the factors the third factor is $(3x+1)$ as shown below.

$15x^3 - 31x^2 + 0x + 4$

$5x - 2$

$\quad x - 2 = 5x^2 - 12x + 4$

$\qquad\qquad 3x + 1 = 15x^3 - 31x^2 + 0x + 4$

$x_1 = 2/5, \quad x_2 = 2, \quad x_3 = -1/3$

This example illustrates perhaps better the origin of the equation as shown below.

$15x^3 - 31x^2 + 0x + 4$

$\dfrac{15}{15}x^3 - \dfrac{31}{15}x^2 + \dfrac{0}{15}x + \dfrac{4}{15}$

$\quad x^3 - \dfrac{31}{15}x^2 + 0x + \dfrac{4}{15}$

$= -2 - \dfrac{1}{15}$

$= -2 - \dfrac{1}{5 \cdot 3}$

$= -2 - \left(\dfrac{2}{5} - \dfrac{1}{3}\right)$

$= -2 - \dfrac{2}{5} + \dfrac{1}{3}$

$\qquad$ **2,07** $\qquad\qquad$ **0,26**

When the answer is multiplied, $2/5 \cdot 2 \cdot 1/3$, we get 0,26 corresponding 4/15 in the last term, and when we add the answer $2/5 + 2 - 1/3$ we get 2,07 corresponding to 31/15 in the coefficient of the x^2-term.

Example 31 $12x^3 + 8x^2 - 3x - 2 = 0$

The left side of the equation is factorised and $12x^3$ is split into $3x/2x/2x$ and 2 into $2/1/1$. Since neither 1 nor minus 1 is a root the factor rule is used. The largest coefficient 3 in the first term is multiplied by the largest factor in the last term. A glance at $+8x^2$ shows that the first factor is probably $(3x+2)$ and the second factor $(2x-1)$. The third factor is given when the two first factors were multiplied as shown below.

$12x^3 + 8x^2 - 3x - 2$

$3x + 2$

$2x - 1 = 6x^2 + x - 2$

$\qquad\qquad 2x + 1 = 12x^3 + 8x^2 - 3x - 2$

$x_1 = -2/3, \ x_2 = 1/2, \ x_3 = -1/2$

Example 32 $12x^3 - 28x^2 - 3x + 7 = 0$

The left side of the equation is factorised and $12x^3$ is split to $3x/2x/2x$ and 7 to $7/1/1$, as neither 1 or minus 1 is a factor. The first factor we get when the largest coefficient 3 in the first term, is multiplied by the largest factor 7 in the last term. A glance at $-28x^2$ tells us that the first factor is likely $(3x-7)$ and the second one $(2x+1)$. After multiplication of the factors the third factor $(2x-1)$ is given as shown below.

$12x^3 - 28x^2 - 3x + 7$

$3x - 7$

$2x + 1 = 6x^2 - 11x - 7$

$\qquad\qquad 2x - 1 = 12x^3 - 28x^2 - 3x + 7$

$x_1 = 7/3, \ x_2 = -1/2, \ x_3 = 1/2$

Example 33 $2x^3 + 3x^2 - 11x - 6 = 0$

The left side of the equation is factorised and $2x^3$ is split into $2x/x/x$ and 6 is split $1/2/3$. After a test the factors are $(x - 2)(x + 3)$. The third factor is $(2x+1)$ as shown below.

$2x^3 + 3x^2 - 11x - 6$

$x - 2$

$x + 3 = x^2 + x - 6$

$\qquad\qquad 2x + 1 = 2x^3 + 3x^2 - 11x - 6$

$x_1 = 2, \ x_2 = -3, \ x_3 = -1/2$

Example 34 $2x^3 - 5x^2 - 23x - 10 = 0$
The left side of the equation is factorised and $2x^3$ is split into 2x/x/x and 10 into 5/2/1.
After a test the factors are $(x - 5)(x + 2)$ and the third factor is $(2x+1)$ as shown below.
$2x^3 - 5x^2 - 23x - 10$
$x - 5$
$x + 2 = x^2 - 3x - 10$
$\qquad 2x + 1 = 2x^3 - 5x^2 - 23x - 10$
$x_1 = 5, \quad x_2 = -2 \quad x_3 = -1/2$

Example 35 $2x^3 + 9x^2 + 3x - 4 = 0$
The left side of the equation is factorised and $2x^3$ is split into 2x/x/x and 4 is split 4/1/1.
After a test the factors are $(x + 4)(x + 1)$ and the third factor is $(2x - 1)$ as shown below.
$2x^3 + 9x^2 + 3x - 4$
$x + 4$
$x + 1 = x^2 + 5x + 4$
$\qquad 2x - 1 = 2x^3 + 9x^2 + 3x - 4$
$x_1 = -4, x_2 = -1, \quad x_3 = 1/2$

Example 36 $4x^3 + x^2 - 4x - 1 = 0$
The left side of the equation is factorised and $4x^3$ is split into 4x/x/x and 1 is split 1/1/1.
After the test the factors are $(x + 1)(x - 1)$. The third factor is $(4x+1)$ as shown below.
$4x^3 + x^2 - 4x - 1$
$x + 1$
$x - 1 = x^2 - 0x - 1$
$\qquad 4x + 1 = 4x^3 + x^2 - 4x - 1$
$x_1 = -1, \quad x_2 = 1, x_3 = -1/4$

Example 37 $6x^3 - 5x^2 - 17x + 6 = 0$

The left side of the equation is factorised and $6x^3$ is split into $3x/2x/x$ and 6 into 3/2/1. After the test $(x - 2)$ is a factor. A glance at $-5x^2$ indicates that the second factor should have a minus sign. We get the second factor when multiplying the largest factor 3, in the first term by 1 in the last term. The second factor should be $(3x - 1)$.

NB! In this example $3x$ cannot form a factor with the largest factor 3 in the last term, as $(3x - 3)$ is equal to $(x - 1)$ and nor with $(x - 2)$. When the factors $(3x - 1)$ och $(x - 2)$ were multiplied the third factor is $(2x+3)$ as shown below.

$6x^3 - 5x^2 - 17x + 6$

$3x - 1$

$\quad x - 2 = 3x^2 - 7x + 2$

$\qquad\qquad 2x + 3 = 6x^3 - 5x^2 - 17x + 6$

$x_1 = 1/3, \ x_2 = 2, x_3 = -3/2$

Example 38 $7x^3 + 13x^2 + 5x - 1 = 0$

The left side of the equation is factorised and $7x^3$ is split into $7x/x/x$ and 1 is split 1/1/1. After a test the factors are $(x + 1)(x + 1)$ and the third factor is $(7x - 1)$ as shown below.

$7x^3 + 13x^2 + 5x - 1$

$x + 1$

$x + 1 = x^2 + 2x + 1$

$\qquad\qquad 7x - 1 = 7x^3 + 13x^2 + 5x - 1$

$x_1 = -1, \ x_2 = -1, \ x_3 = 1/7$

Example 39 $5x^3 - 9^2 + 3x + 1 = 0$

The left side of the equation is factorised and $5x^3$ is split into $5x/x/x$ and 1 is split 1/1/1. After a test both factors are $(x - 1)(x - 1)$ and the third factor is $(5x+1)$ as shown below.

$5x^3 - 9x^2 + 3x + 1$

$x - 1$

$x - 1 = x^2 - 2x + 1$

$\qquad\qquad 5x + 1 = 5x^3 - 9x^2 + 3x + 1$

$x_1 = 1, \ x_2 = 1, \ x_3 = -1/5$

Example 40 $6x^3 - 5x^2 - 16x + 15 = 0$

The left side of the equation is factorised and $6x^3$ is split into $3x/2x/x$ and 15 into 5/3/1. The first factor is $(x-1)$. The second factor is obtained when the largest coefficient 3 in the first term, is multiplied by the largest factor 5 in the last term. After a test the second factor is $(3x+5)$ and the third factor $(2x - 3)$ is given as shown below.

$6x^3 - 5x^2 - 16x + 15$
$3x + 5$
$\quad x - 1 = 3x^2 + 2x - 5$
$\qquad\qquad 2x - 3 = 6x^3 - 5x^2 - 16x + 15$
$x_1 = -5/3, \ x_2 = 1, \ x_3 = 3/2$

Example 41 $3x^3 + 5x^2 - 3x - 5 = 0$

The left side of the equation is factorised and $3x^3$ is split into $3x/x/x$ and 5 is split 5/1/1. The factors are $(x - 1)(x + 1)$ and the third factor $(3x+5)$ is given as shown below.

$3x^3 + 5x^2 - 3x - 5$
$x - 1$
$x + 1 = x^2 + 0x - 1$
$\qquad\qquad 3x + 5 = 3x^3 + 5x^2 - 3x - 5$
$x_1 = 1, \ x_2 = -1, \ x_3 = -5/3$

Example 42 $10x^3 + 11x^2 - 2x - 3 = 0$

The left side of the equation is factorised and $10x^3$ is split into $5x/2x/x$ and 3 into 3/1/1. The first factor is $(x+1)$. The second factor is obtained when the largest coefficient $5x$ in the first term is multiplied by the largest factor 3 in the last term. A glance at $+11x^2$ tells us that the factor should be $(5x+3)$. After multiplication of the factors the third factor is $(2x-1)$ as shown below.

$10x^3 + 11x^2 - 2x - 3$
$\quad 5x + 3$
$\qquad x + 1 = 5x^2 + 8x + 3$
$\qquad\qquad 2x - 1 = 10x^3 + 11x^2 - 2x - 3$
$x_1 = -3/5, \ x_2 = -1, \ x_3 = 1/2$

Example 43 $4x^3 + 8x^2 + x - 3 = 0$

The left side of the equation is factorised and $4x^3$ is split into 2x/2x/x and 3 into 3/1/1. The first factor is $(x + 1)$. The second factor is obtained when the largest coefficient 2 in the first term, is multiplied by the largest factor 3 in the last term. A glance at $+8x^2$ tells us that the second factor should be $(2x + 3)$. When the factors were multiplied, the third factor $(2x-1)$ is given as shown below.

$4x^3 + 8x^2 + x - 3$

$2x + 3$

$\quad x + 1 = 2x^2 + 5x + 3$

$\qquad\qquad 2x - 1 = 4x^3 + 8x^2 + x - 3$

$x_1 = -3/2, \quad x_2 = -1, \quad x_3 = 1/2$

Example 44 $7x^3 - 34x^2 + 37x + 6 = 0$

The left side of the equation is factorised and $7x^3$ is split into 7x/x/x and 6 is split 3/2/1. After a test the factors are $(x - 2)(x - 3)$ and after multiplication the third factor $(7x + 1)$ is given as shown below.

$7x^3 - 34x^2 + 37x + 6$

$x - 2$

$x - 3 = x^2 - 5x + 6$

$\qquad\qquad 7x + 1 = 7x^3 - 34x^2 + 37x + 6$

$x_1 = 2, \quad x_2 = 3 \quad x_3 = -1/7$

Example 45 $5x^3 + 2x^2 - 5x - 2 = 0$

The left side of the equation is factorised and $5x^3$ is split into 5x/x/x and 2 is split 2/1/1. The factors are $(x + 1)(x - 1)$ and the third factor $(5x + 2)$ is given as shown below.

$5x^3 + 2x^2 - 5x - 2$

$x + 1$

$x - 1 = x^2 + 0x - 1$

$\qquad\qquad 5x + 2 = 5x^3 + 2x^2 - 5x - 2$

$x_1 = -1, \quad x_2 = 1, \quad x_3 = -2/5$

Example 46 $3x^3 + 2x^2 - 12x - 8 = 0$
The left side of the equation is factorised and $3x^3$ is split into $3x/x/x$ and 8 is split 2/2/2,
since $(x + 2)(x - 2)$ are factors. The third factor $(3x+2)$ is given as shown below.
$3x^3 + 2x^2 - 12x - 8$
  $x + 2$
  $x - 2 = x^2 + 0x - 4$
            $3x + 2 = 3x^3 + 2x^2 - 12x - 8$
$x_1 = -2$, $x_2 = 2$, $x_3 = -2/3$

Example 47 $2x^3 + 5x^2 - 4x - 3 = 0$
The left side of the equation is factorised and $2x^3$ is split into $2x/x/x$ and 3 is split 3/1/1.
After a test $(x - 1)(x + 3)$ are factors and the third factor is $(2x+1)$ as shown below.
$2x^3 + 5x^2 - 4x - 3$
$x - 1$
$x + 3 = x^2 + 2x - 3$
            $2x + 1 = 2x^3 + 5x^2 - 4x - 3$
$x_1 = 1$, $x_2 = -3$, $x_3 = - 1/2$

Example 48 $2x^3 + x^2 - 13x + 6 = 0$
The left side of the equation is factorised and $2x^3$ is split into $2x/x/x$ and 6 is split 3/2/1.
After a test $(x-2)(x+3)$ are factors and the third factor is $(2x-1)$ as shown below.
$2x^3 + x^2 - 13x + 6$
$x - 2$
$x + 3 = x^2 + x - 6$
            $2x - 1 = 2x^3 + x^2 - 13x + 6$
$x_1 = 2$, $x_2 = -3$, $x_3 = 1/2$

Example 49 $3x^3 + 4x^2 - 5x - 2 = 0$
The left side of the equation is factorised and $3x^3$ is split into $3x/x/x$ and 2 is split 2/1/1.
After a test the factors are $(x -1)(x + 2)$ and the third factor is $(3x +1)$ as shown below.
$3x^3 + 4x^2 - 5x - 2$
$x - 1$
$x + 2 = x^2 + x - 2$
            $3x + 1 = 3x^3 + 4x^2 - 5x - 2$
$x_1 = 1$, $x_2 = -2$, $x_3 = - 1/3$

Example 50 $2x^3 + 4x^2 - 2x - 4 = 0$
The left side of the equation is factorised and $2x^3$ is split into $2x/x/x$ and 4 is split $2/2/1$. After a test $(x + 1)(x + 2)$ are factors. The third factor $(2x - 2)$ is given as shown below.
$2x^3 + 4x^2 - 2x - 4$
$x + 1$
$x + 2 = x^2 + 3x + 2$
$\qquad\qquad 2x - 2 = 2x^3 + 4x^2 - 2x - 4$
$x_1 = -1, \quad x_2 = -2, \quad x_3 = 1$

Example 51 $3x^3 + x^2 - 12x - 4 = 0$
The left side of the equation is factorised and $3x^3$ is split into $3x/x/x$ and 4 is split $1/2/2$. After a test $(x + 2)(x - 2)$ are factors. The third factor is $(3x + 1)$ as shown below.
$3x^3 + x^2 - 12x - 4$
$x + 2$
$x - 2 = x^2 + 0x - 4$
$\qquad\qquad 3x + 1 = 3x^3 + x^2 - 12x - 4$
$x_1 = -2, \quad x_2 = 2, \quad x_3 = -1/3$

Example 52 $4x^3 - x^2 - 36x + 9 = 0$
The left side of the equation is factorised and $4x^3$ is split into $4x/x/x$ and 9 is split $3/3/1$. After a test $(x + 3)(x - 3)$ are factors. The third factor is $(4x - 1)$ as shown below.
$4x^3 - x^2 - 36x + 9$
$x + 3$
$x - 3 = x^2 - 0x - 9$
$\qquad\qquad 4x - 1 = 4x^3 - x^2 - 36x + 9$
$x_1 = -3, \quad x_2 = 3, \quad x_3 = 1/4$

Example 53 $2x^3 - 3x^2 - 5x + 6 = 0$
The left side of the equation is factorised and $2x^3$ is split into $2x/x/x$ and 6 is split $3/2/1$. After a test $(x - 1)(x - 2)$ are factors. The third factor is $(2x + 3)$ as shown below.
$2x^3 - 3x^2 - 5x + 6$
$x - 1$
$x - 2 = x^2 - 3x + 2$
$\qquad\qquad 2x + 3 = 2x^3 - 3x^2 - 5x + 6$
$x_1 = 1, \quad x_2 = 2, \quad x_3 = -3/2$

Example 54 $3x^3 - x^2 - 10x + 8 = 0$

The left side of the equation is factorised and $3x^3$ is split into $3x/x/x$ and 8 into 4/2/1.
After a test the factors are $(x - 1)(x + 2)$. The third factor is $(3x - 4)$ as shown below.

$3x^3 - x^2 - 10x + 8$

$x - 1$

$x + 2 = x^2 + x - 2$

$\qquad\qquad 3x - 4 = 3x^3 - x^2 - 10x + 8$

$x_1 = 1, \quad x_2 = -2, \quad x_3 = 4/3$

Example 55 $2x^3 - 9x^2 + x + 12 = 0$

The left side of the equation is factorised and $2x^3$ is split into $2x/x/x$ and 12 into 4/3/1.
After a test $(x + 1)(x - 4)$ are factors. The third factor is $(2x - 3)$ as shown below.

$2x^3 - 9x^2 + x + 12$

$x + 1$

$x - 4 = x^2 - 3x - 4$

$\qquad\qquad 2x - 3 = 2x^3 - 9x^2 + x + 12$

$x_1 = -1, \quad x_2 = 4, \quad x_3 = 3/2$

Use the following method

Multiply the largest factor in the first term by the factors in the last term.

When the product is less than the coefficient of the x-term, the largest factor in the first term shall be part of the binomial together with smallest factor in the last term.

But when the product is larger than the coefficient of the x-term, the largest factor in the first term shall be part of the binomial with the middle factor in the last term.

When the middle factor in the first term is larger than 1, it shall be part of the binomial together with the largest factor in the last term.

Example 56 $5x^3 + 31x^2 + 51x + 9 = 0$

The left side of the equation is factorised and $5x^3$ is split to $5x/x/x$ and 9 is split to $1/3/3$. Multiply the first and largest coefficient $5x$ by the factors in the last term and we get:
$5x \cdot 1 \cdot 3 \cdot 3 = 45x < 51x$. That means the first factor is $(5x+1)$ and the other two factors are $(x + 3)(x + 3)$ as shown below.
$5x^3 + 31x^2 + 51x + 9$
$5x + 1$
$\quad x + 3 = 5x^2 + 16x + 3$
$\qquad\qquad\quad x + 3 = 5x^3 + 31x^2 + 51x + 9$
$x_1 = -1/5,\ x_2 = x_3 = -3$

Example 57 $5x^3 + 42x^2 + 51x + 14 = 0$
The left side of the equation is factorised and $5x^3$ is split to $5x/x/x$ and 14 is split $1/2/7$. Multiply the first and largest coefficient $5x$ by the factors in the last term and we get:
$5x \cdot 1 \cdot 2 \cdot 7 = 70x > 51x$. That means the first factor is $(5x+2)$. The other two factors are $(x + 7)\,(x + 1)$ as shown below.
$5x^3 + 42x^2 + 51x + 14$
$5x + 2$
$\quad x + 7 = 5x^2 + 37x + 14$
$\qquad\qquad\quad x + 1 = 5x^3 + 42x^2 + 51x + 14$
$x_1 = -2/5,\ x_2 = -7,\ x_3 = -1$

Example 58 $8x^3 + 58x^2 + 110x +24 = 0$

The left side of the equation is factorised and $8x^3$ is split into 4x/2x/x and 24 into 1/4/6. Multiply the first and largest coefficient 4x by the factors in the last term and we get: $4x \cdot 1 \cdot 4 \cdot 6 = 96x < 110x$. The first factor is $(4x+1)$ and the other two factors are $(2x+6)$ and $(x + 4)$ as shown below.

$8x^3 + 58x^2 + 110x + 24$

$4x + 1$

$2x + 6 = 8x^2 + 26x + 6$

$\qquad\qquad x + 4 = 8x^3 + 58x^2 + 110x + 24$

$x_1 = -1/4,\ x_2 = -3,\ x_3 = -4$

Example 59 $7x^3 + 85x^2 + 201x +27 = 0$

The left side of the equation is factorised and $7x^3$ is split to 7x/x/x and 27 is split 1/3/9. Multiply the first and largest coefficient 7x by the factors in the last term and we get: $7x \cdot 1 \cdot 3 \cdot 9 = 189x < 201$.The first factor is $(7x+1)$ and the other two factors are $(x+3)$ and $(x + 9)$ as shown below.

$7x^3 + 85x^2 + 201x +27$

$7x + 1$

$\quad x + 3 = 7x^2 + 22x + 3$

$\qquad\qquad x + 9 = 7x^3 + 85x^2 + 201x + 27$

$x_1 = -1/7,\ x_2 = -3,\ x_3 = -9$

Example 60 $3x^3 + 17x^2 + 22x +8 = 0$

The left side of the equation is factorised and $3x^3$ is split into 3x/x/x and 8 is split 1/2/4. Multiply the first and largest coefficient 3x by the factors in the last term and we get: $3x \cdot 1 \cdot 2 \cdot 4 = 24x > 22x$ and the first factor is $(3x+2)$. The other two factors are $(x + 4)$ $(x + 1)$ as shown below.

$3x^3 + 17x^2 + 22x +8$

$3x + 2$

$\quad x + 4 = 3x^2 + 14x + 8$

$\qquad\qquad x + 1 = 3x^3 + 17x^2 + 22x + 8$

$x_1 = -2/3,\ x_2 = -4,\ x_3 = -1$

Prime factorising

Prime factorising is often used for large numbers. This means that an integer is rewritten as a product of prime numbers, e.g. $15 = 3 \cdot 5$.

A prime number is a positive integer and is only divisible by 1and itself e. g. 2, 3, 5, etc. Integers that are not prime numbers are called composite numbers. They can be split to two or more prime numbers, whose product is the number itself. They are called prime factors, e.g. $10 = 2 \cdot 5$.

When determining the prime factors of e.g. 24, the number is written as a product of the factors 2 and 12. The number 12 can be written as a product of 2 and 6 and the number 6 can be written as a product of 2 and 3 as shown.

$24 = 2 \cdot 12 = 2 \cdot 2 \cdot 6 = 2 \cdot 2 \cdot 2 \cdot 3$

Number 24 can also be written as a product of 4 and 6, since $4 = 2 \cdot 2$ and $6 = 2 \cdot 3$.
$24 = 4 \cdot 6 = 2 \cdot 2 \cdot 2 \cdot 3$

By prime factorising of e.g. 36, it does not matter whether 18 is written as a product of 2 and 9 or 3 and 6 as shown.

$36 = 2 \cdot 18 = 2 \cdot 2 \cdot 9 = 2 \cdot 2 \cdot 3 \cdot 3$ or $36 = 2 \cdot 18 = 2 \cdot 3 \cdot 6 = 2 \cdot 2 \cdot 3 \cdot 3$

Prime factorising is also used to calculate the greatest common divider of two or more integers and it will be well suited by fraction. When the numerator and the denominator have a factor in common, the fraction is cancelled. In the following example the numbers, 70 and 42, have the prime factors 2 and 7 in common. It means that the largest common dividend to 70 and 42 is 2 and 7.
$70 = 2 \cdot 35 = 2 \cdot 5 \cdot 7$ and $42 = 2 \cdot 21 = 2 \cdot 3 \cdot 7$

$$\frac{70}{42} = \frac{2 \cdot 5 \cdot 7}{2 \cdot 3 \cdot 7} = \frac{5}{3}$$

The example shows that prime factors, which are found in the numerator and the denominator, can be cancelled. The product is the largest common dividend i.e. $2 \cdot 2 \cdot 5 = 20$.

$180 = 2 \cdot 90 = 2 \cdot 2 \cdot 45 = 2 \cdot 2 \cdot 9 \cdot 5 = 2 \cdot 2 \cdot 3 \cdot 3 \cdot 5$ and
$280 = 2 \cdot 140 = 2 \cdot 2 \cdot 70 = 2 \cdot 2 \cdot 2 \cdot 35 = 2 \cdot 2 \cdot 2 \cdot 5 \cdot 7$

$$\frac{180}{280} = \frac{2 \cdot 2 \cdot 3 \cdot 3 \cdot 5}{2 \cdot 2 \cdot 2 \cdot 5 \cdot 7} = \frac{3 \cdot 3}{2 \cdot 7}$$

Another application is to simplify expressions of roots as shown.

$$180 = 2 \cdot 90 = 2 \cdot 2 \cdot 45 = 2 \cdot 2 \cdot 5 \cdot 9 = 2 \cdot 2 \cdot 3 \cdot 3 \cdot 5$$

$$\sqrt{180} = \sqrt{2 \cdot 2 \cdot 3 \cdot 3 \cdot 5} = \sqrt{2 \cdot 2 \cdot 3 \cdot 3} \cdot \sqrt{5} = 2 \cdot 3 \cdot \sqrt{5}.$$

However, prime factorising is most common by encryption of passwords, where large prime numbers are used to make a log in as safe as possible, i.e. when we log in to our internet banking. A common application of prime numbers is for encryption, where two very large prime numbers are multiplied. It is very difficult to find the prime factors, as you only know the product and that makes our Internet bank safe.

Chapter 4

FOURTH DEGREE EQUATIONS

Example 1 $3x^4 + 26x^3 + 79x^2 + 96x + 36 = 0$

The left side of the equation is factorised and $3x^4$ is split into $3x/x/x/x$ and 36 is split by prime factorising: $36 = 2 \cdot 18 = 2 \cdot 2 \cdot 9 = 2 \cdot 2 \cdot 3 \cdot 3$.

When we multiply the first and largest coefficient 3, with the factors in the last term, we get; $3x \cdot 2 \cdot 2 \cdot 3 \cdot 3 = 108x > 96x$. The first factor is $(3x+2)$ and the other three factors are $(x + 2)(x+ 3)(x + 3)$ as shown below.

$3x^4 + 26x^3 + 79x^2 + 96x + 36$

$3x + 2$

   $x + 2 = 3x^2 + 8x + 4$

                  $x + 3 = 3x^3 + 17x^2 + 28x + 12$

                                 $x + 3 = 3x^4 + 26x^3 + 79x^2 + 96x + 36$

$x_1 = -2/3, \quad x_2 = -2, \quad x_3 = x_4 = -3$

Example 2 $x^4 + 10x^3 + 37x^2 + 60x + 36 = 0$ $(2+2+3+3 = 10x^3)$

The left side of the equation is factorised and $3x^4$ is split into $x/x/x/x$ and 36 is split by prime factorising: $36 = 2 \cdot 18 = 2 \cdot 2 \cdot 9 = 2 \cdot 2 \cdot 3 \cdot 3$.

Since the first coefficient is 1, the sum of the factors is the same as the coefficient of the x^3-term 10 and the product is 36. The factors are $(x + 2)(x + 2)(x + 3)(x + 3)$ as shown below.

$x^4 + 10x^3 + 37x^2 + 60x + 36$

$x + 2$

$x + 2 = x^2 + 4x + 4$

            $x + 3 = x^3 + 7x^2 + 16x + 12$

                           $x + 3 = x^4 + 10x^3 + 37x^2 + 60x + 36$

$x_1 = x_2 = -2, \quad x_3 = x_4 = -3.$

Comparison with a conventional method

Substitution

$x^4 + 10x^3 + 37 x^2 + 60x + 36 = 0$

The equation is broken down to a quadratic equation by substitution, which means that parts of the equation are replaced by an expression, where x^4 and $10x^3$ is included. In order to get the right expression, we halve the coefficient 10 in the variable term x^3 and multiply x^2 by 5x. By cross-multiplying we get as shown.

$x^2 + 5x$

$x^2 + 5x = x^4 + 10x^3 + 25x^2$

We put $t = (x^2 + 5x)$ and $t^2 = (x^2 + 5x)^2$

$x^4 + 10 x^3 + 37 x^2 + 60x + 36 = 0$

$(x^2 + 5x)^2 = x^4 + 10x^3 + 25x^2$

$x^4 + 10x^3 + 25x^2 + 12x^2 + 60x + 36 = 0$

$(x^2 + 5x)^2 + 12 (x^2 + 5x) + 36 = 0$

$t^2 + 12t + 36 = 0$

Solve the quadratic equation: $t^2 + 12t + 36 = 0$

$t = -6 \pm \sqrt{36 + 1(-36)}$

$t = -6$, put $t = x^2 + 5x$

$x^2 + 5x + 6 = 0$

$x = -2{,}5 \pm \sqrt{6{,}25 + 1(-6)} = -2{,}5 \pm \sqrt{0{,}25} = -2{,}5 \pm 0{,}5$

$\therefore x_1 = x_2 = -2, \ x_3 = x_4 = -3.$

Chapter 5

FIFTH DEGREE EQUATIONS

Example 1 $5x^5 + 67x^4 + 163x^3 + 253x^2 + 168x + 36 = 0$

The left side of the equation is factorised and $5x^5$ is split into $5x/x/x/x/x$ and by prime factorising of the constant term 36 we get: $36 = 2 \cdot 18 = 2 \cdot 2 \cdot 9 = 2 \cdot 2 \cdot 3 \cdot 3$.
Since we have five factors and the product must be 36, the fifth factor has to be 1. The factors are accordingly 1, 2, 2, 3, 3, and the product is 36.

Multiply the first and the largest coefficient $5x$ in the first term by the factors in the last term and we get: $5x \cdot 2 \cdot 2 \cdot 3 \cdot 3 \cdot 1 = 180x > 168x$ and the first factor is $(5x+2)$. We get the factors $(5x+2)(x + 1)(x + 2)(x +3)(x + 3)$ as shown below.

$5x^5 + 47x^4 + 163x^3 + 253x^2 + 168x + 36$
$5x +2$
  $x + 1 = 5x^2 + 7x + 2$
            $x + 2 = 5x^3 + 17x^2 + 16x + 4$
                      $x + 3 = 5x^4 + 32x^3 + 67x^2 + 52x + 12$
                                $x + 3 = 5x^5 + 47x^4 + 163x^3 + 253x^2 + 168x + 36$

$x_1 = -2/5, \ x_2 = -1, \ x_3 = -2, \ x_4 = x_5 = -3.$

Example 2 $x^5 + 11x^4 + 47x^3 + 97x^2 + 96x + 36 = 0$ $(1 + 2 + 2 + 3 + 3 = 11x^4)$

The left side of the equation is factorised and x^5 is split into $x/x/x/x/x$ and by means of prime factorising of the constant term we get: $36 = 2 \cdot 18 = 2 \cdot 2 \cdot 9 = 2 \cdot 2 \cdot 3 \cdot 3$.
Since we have five factors and the product has to be 36, the fifth factor has to be 1. So the sum of the factors is 11 like the coefficient of the x^4-term and the product is 36.

The coefficient of the x^4-term contains all information about the origin of the equation. It can be compared to a cell nucleus, containing the whole genetic material i.e. its DNA. The five factors are $(x + 1)(x + 2)(x + 2)(x + 3)(x + 3)$ as shown below.

$x^5 + 11x^4 + 47x^3 + 97x^2 + 96x + 36$
$x + 1$
$x + 2 = x^2 + 3x + 2$
          $x + 2 = x^3 + 5x^2 + 8x + 4$
                    $x + 3 = x^4 + 8x^3 + 23x^2 + 28x + 12$
                              $x + 3 = x^5 + 11x^4 + 47x^3 + 97x^2 + 96x + 36$

$x_1 = -1, \ x_2 = x_3 = -2, \ x_4 = x_5 = -3$

The graph to the fifth degree function: $y = x^5 + 11x^4 + 47x^3 + 97x^2 + 96x + 36$

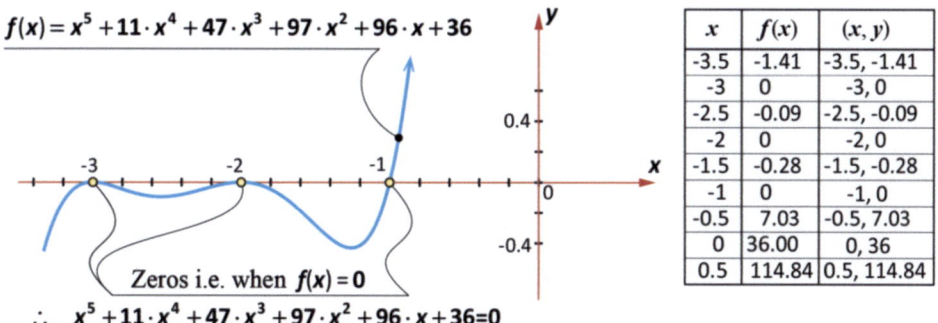

$f(x) = x^5 + 11 \cdot x^4 + 47 \cdot x^3 + 97 \cdot x^2 + 96 \cdot x + 36$

x	$f(x)$	(x, y)
-3.5	-1.41	-3.5, -1.41
-3	0	-3, 0
-2.5	-0.09	-2.5, -0.09
-2	0	-2, 0
-1.5	-0.28	-1.5, -0.28
-1	0	-1, 0
-0.5	7.03	-0.5, 7.03
0	36.00	0, 36
0.5	114.84	0.5, 114.84

Zeros i.e. when $f(x) = 0$

$\therefore \quad x^5 + 11 \cdot x^4 + 47 \cdot x^3 + 97 \cdot x^2 + 96 \cdot x + 36 = 0$

We compare with a conventional method.

Polynomial division

Polynomial division presume that you in some way know the root, or if nothing else, by a guess or by a test. In this example the root is $x = -1$.

The conventional way of solving a fifth degree equation is carried out in three steps: Polynomial division, substitution and a quadratic equation solved by a formula.

By polynomial division with $(x + 1)$, the equation is broken down to a fourth degree equation and through substitution the fourth degree equation is broken down to a quadratic equation.

$x^5 + 11x^4 + 47x^3 + 97x^2 + 96x + 36 = 0$

$x^4 + 10x^3 + 37x^2 + 60x + 36$
$\overline{x^5 + 11x^4 + 47x^3 + 97x^2 + 96x + 36} / x + 1$
$\underline{x^5 + x^4} \qquad\qquad\qquad\qquad\qquad x^4 \ (x+1)$
$\quad 10x^4 + 47x^3 \qquad\qquad\qquad\quad 10x^3 \ (x+1)$
$\quad \underline{10x^4 + 10x^3} \qquad\qquad\qquad\quad 37x^2 \ (x+1)$
$\qquad\quad 37x^3 + 97x^2 \qquad\qquad\quad 60x \ (x+1)$
$\qquad\quad \underline{37x^3 + 37x^2} \qquad\qquad\quad 36 \ (x+1)$
$\qquad\qquad\quad 60x^2 + 96x$
$\qquad\qquad\quad \underline{60x^2 + 60x}$
$\qquad\qquad\qquad\quad 36x + 36$
$\qquad\qquad\qquad\qquad\quad 0$

87

Substitution

$x^4 + 10x^3 + 37x^2 + 60x + 36 = 0$

By substitution this fourth degree equation is broken down to a quadratic equation, which means that a part of the equation is replaced by an expression containing x^4 and $10x^3$. Cross-multiply $x^2 + 5x$ as shown.

$x^2 + 5x$

$x^2 + 5x = x^4 + 10x^3 + 25x^2$ i.e. $(x^2 + 5x)^2 = x^4 + 10x^3 + 25x^2$

We choose to put $t = (x^2 + 5x)$ and $t^2 = (x^2 + 5x)^2$.

$x^4 + 10x^3 + 37x^2 + 60x + 36 = 0$
$(x^2 + 5x)^2 = x^4 + 10x^3 + 25x^2$
$x^4 + 10x^3 + 25x^2 + 12x^2 + 60x + 36 = 0$
$(x^2 + 5x)^2 + 12(x^2 + 5x) + 36 = 0$
$t^2 + 12t + 36 = 0$

Solve the equation $t^2 + 12t + 36 = 0$

$t = -6 \pm \sqrt{36 + 1(-36)}$

$t = -6$ put $t = x^2 + 5x$

$x^2 + 5x + 6 = 0$

$x = -2{,}5 \pm \sqrt{6{,}25 + 1(-6)} = -2{,}5 \pm \sqrt{0{,}25} = -2{,}5 \pm 0{,}5$

$x_1 = -1, x_2 = x_3 = -2, \ x_4 = x_5 = -3.$

Chapter 6

THE GREATEST MATHEMATICIAN IN NORTHERN EUROPE

Niels Henrik Abel (5:e Aug 1802 – 6:e April 1829)

Niels Henrik Abel was a son o f the pastor in Gjerstad near Risör in Norway. Abel has been described as the greatest mathematician in the Nordic countries. He is perhaps best known for his complete proof of demonstrating the impossibility of solving a fifth degree equation with algebraic solutions. Although he died at the age of only 26, he made pioneering contributions in a variety of fields. Several of them became known only after his death and one of them, and probably one of the greatest theses in the history of mathematics, is the one he wrote in Paris in 1824. He wrote about integrals and elliptic functions, where he showed the connection between algebra, mathematical analysis and geometry, which no one had discovered before.

The thesis disappeared several times and in the end it was found in Florence, Italy in 1952. This thesis is now protected at the University of Oslo.

At the age of 13 he entered the Cathedral School in Oslo, former called Christiania. Here he discovered mathematics and acquired new knowledge with such an eagerness that his teacher wrote in his report :"An excellent mathematical genius".

When Abel entered the University of Oslo in 1821, he was already the best mathematician in Norway. Professor Holmboe was his teacher and had no more to learn him. Abel studied the latest mathematical literature at the university library and that time he started working on the fifth degree equation in radicals. Mathematicians had been looking for a solution to this problem for more than 250 years and Abel thought he had found the solution.

The professors in mathematics in Oslo, Christophe Rasmussen and Søren Rasmussen, found no error in Abel's formulas, and therefore they sent his work to the leading mathematician, Professor Degen in Copenhagen. He found no faults either, but still he doubted that the solution, which many outstanding mathematicians had sought for so long, had been discovered by an unknown student from Norway.

Professor Degen realized Abel's sharp mind and suggested that such a young man should not waste time on "a sterile object" as the fifth degree equation. He suggested that Abel instead should work with elliptic functions and transcendence. The theorem for the elliptic functions has been developed mainly by Niels Henrik Abel.

Even though no one could find anything wrong with the formula, Abel himself discovered that this formula could not be general for all fifth degree equations. At that time Abel did not know that Paolo Ruffini from Italy had left a proof about 25 years earlier, but afterwards he found however, that neither his own proof nor Ruffini's was tenable. Later he showed two complete proofs in this regard and the statement is today called the Abel-Ruffini theorem.

Abel's life was dominated by poverty and he had to rely on donations from friends. His father passed away in 1820 and his mother was sickly and could not take care of her children and her home. His financial situation however, was improved when he met Christine Kemp in Copenhagen, who later became his wife and financier.

Niels Henrik Abel died of tuberculosis at the age of 26. Two days after his death a letter from Berlin arrived, stating that he had been offered a professorship at the University of Berlin, but the good news came too late.

At that time almost no one understood Abel's greatness, but today many people realize how important he was for the science of mathematics.

Several streets in Norway as well as in Berlin and Paris have been named after Abel. He has also been compared to great Norwegian celebrities, such as the composer Edvard Grieg, the artist Edvard Munch and the dramatist Henrik Ibsen.

The Abel prize

The Abel Prize is a Norwegian annual award from the Government of Norway to one or more outstanding mathematicians. The Norwegian Government offered this international prize in mathematics an initial funding of 200 million Norwegian crowns NOK, or about 21.7 million Euro. The prize was for the first time awarded in June 2003 with an award of 6 million Norwegian crowns or around 650.000 Euro.

The Norwegian Academy of Science declares the winner of the Abel Prize each year in March on the recommendation of the Abel Committee, which consists of five leading mathematicians. The International Mathematical Union and the European Mathematical Society nominate the members of the Abel Committee.

The prize is presented to the winner by the King of Norway, and later the day ends with a banquet at the castle of Akershus.

Since there is no Nobel prize in mathematics, the Abel prize is important and has got a high status within mathematics.

References

1. Williams K.R. (1991). The Natural Calculator

2. Williams K.R. (1984). Discover Vedic Mathematics. Inspiration Books

3. Bidder G.P. (1856). On Mental Calculation, Minutes of Proceedings, Institution of Civil Engineers (1855-56), 15, 251-280

4. Aitken A.C. (1954). The Art of Mental Calculation: With Demonstrations. Transactions of the Society of Engineers. 45, 295-309

5. Tirthaji B.K. (1965). Vedic Mathematics, Motilal Banarsidass

6. Nicholas, Williams, Pickles (1984). Vertically and Crosswise. Inspiration Books

7. Stubhaug, A. Ett foranskutt lyn. Niels Henrik Abel and his time.